Probability Theory

Key Concepts and Tools
for
SOA Exam P & CAS Exam 1

Olivier Le Courtois, PhD, FSA, CFA, CERA, FRM

Probability Theory.
Key Concepts and Tools for SOA Exam P & CAS Exam 1.

Copyright © 2018 by Olivier Le Courtois.

All rights reserved. No part of this book may be reproduced or transmitted in any form or by any means, electronic or mechanical, including photocopying, recording, or by any information storage and retrieval system without the written permission of the publisher, except where permitted by law.

Cover design by Emiko Muraoka.

Copyediting by Jonathan Moore.

Contents

1 Introduction   1

2 General probability   4
    2.1 Set functions and probability measures   4
    2.2 Mutually exclusive events   9
    2.3 Addition and multiplication rules   10
    2.4 Independence of events   11
    2.5 Combinatorial probability   12
    2.6 Conditional probability   13
    2.7 The law of total probability and Bayes' theorem .   15

3 Univariate probability distributions   17
    3.1 Probability mass and density functions   17
    3.2 Cumulative distribution functions   24
    3.3 Mode, percentiles, median, mean, and moments . .   29
    3.4 Variance and measures of dispersion   40
    3.5 Sums of independent random variables   45
    3.6 Moment generating functions   47
    3.7 Transformations   49

4 Multivariate probability distributions   52
    4.1 Joint and marginal probability functions and densities   52
    4.2 Joint cumulative distribution functions   56
    4.3 Central limit theorem   58
    4.4 Conditional probability distributions   58
    4.5 Moments   59

4.6	Joint moment generating functions	64
4.7	Variance and measures of dispersion	65
4.8	Covariance and correlation coefficients	66
4.9	Transformations and order statistics	69

Appendix   **70**

Chapter 1

Introduction

I wish I had had access to a book like this one when I took the Exam P of the Society of Actuaries (SOA). It contains all the formulas that you need to solve the official exercises of the SOA. I have tried to keep it very compact, theoretically solid, and not verbose.

Students who are not contemplating becoming actuaries but who are more interested in econometrics, finance, statistics, mathematics, or other fields, will also find this text useful.

I also recommend this book as a prerequisite to the students who are considering taking, or are in the process of taking, the CFA exams. Indeed, the statistics and portfolio management material studied in the CFA syllabus are based on the probability results shown in this book.

The order in which the contents of this book are presented mostly respects the order of the SOA and Casualty Actuarial Society (CAS) syllabi. I made very few adjustments to this order and I did so only when pedagogical reasons supported the change.

This book does not just present the material; it helps you understand the foundations of the material. To keep the book compact, I purposely did not include exercises. On the contrary, this book should be used with the (long) series of exercises made freely available by the SOA.

The tables in the appendix link those exercises to the equations in the book. These tables can be a very convenient tool for providing hints for the exercises that you cannot solve – instead of going directly to the solutions.

I also suggest that you use this book in conjunction with the "BTDT (Been There Done That) Study Manual for Exam P/1" written by Prof. Krzysztof Ostaszewski. This book can be obtained from http://smartURL.it/BTDT-P. I also suggest that you take a look at http://smartURL.it/PassExamP for a wealth of exercises with audio explanations.

This book is the first in a series that I am writing for actuarial associateship exams. In each of these books, I provide conceptual links between the contents of the various exams. This book was also written in such a way that you can use it throughout your career.

My best suggestion for those who are taking the exams is to avoid using flashcards. Why? Because if you need to quickly check a few days before an exam that you know your formulas, it just means you do not really know these formulas. More precisely, it means you have not mastered the concepts behind these formulas. So, if you are serious and if you have mastered your textbooks and have done hundreds of exercises, then flashcards are just completely useless because you know your formulas and the concepts behind them.

In the context of actuarial exams, I consider flashcards to be very dangerous for an additional reason: what you learn for an exam will be useful for the subsequent exams. Therefore, a deep mastery of the first exams is important when you start preparing for the most advanced exams. Again, this deep mastery is not attainable using "easy" tricks such as flashcards.

At this time, I want to thank Li Shen for technical assistance. I also want to thank Xia Xu and Xiaoshan Su for pointing out typos.

Before delving into this book, you should make sure that you are comfortable with classic calculus (summation, differentiation, integration, and so on). I will assume throughout the text that you know the following geometric series by heart. When the indexation starts from zero:

$$\sum_{i=0}^{N} k^i = \frac{1 - k^{N+1}}{1 - k} \tag{1}$$

and

$$\sum_{i=0}^{+\infty} k^i = \frac{1}{1 - k}. \tag{2}$$

When the indexation starts from one:

$$\sum_{i=1}^{N} k^i = \frac{k - k^{N+1}}{1 - k} \tag{3}$$

and

$$\sum_{i=1}^{+\infty} k^i = \frac{k}{1 - k}. \tag{4}$$

From now on, I will also assume that you are comfortable with the notation $\max(x, y) = (x, y)^+$ and $\min(x, y) = (x, y)^-$.

To find out information about my publications, to get advice on the SOA and CAS exams, to kindly suggest corrections, or to subscribe to my monthly newsletter, see:

www.olivierlecourtois.com

Chapter 2

General probability

In this chapter, we will study the main concepts of probability theory. The presentation starts by introducing the framework of sets, then it introduces the most classic probability rules and culminates with a presentation of the Bayes theorem.

2.1 Set functions and probability measures

Let Ω be the biggest set available. All the other sets, such as A and B, are subsets of Ω and verify $A \cap \Omega = A$, where $\cap$ is the **intersection** operator. We also denote as $\cup$ the **union** operator.

Figure 1 presents the simplest illustration of the intersection and union operators. In this figure, the set A is the union of the surfaces tagged 1 and 3, and is the disk on the left. The set B is in turn the union of surfaces 2 and 3. Conversely, surface 1 can be expressed as a function of the sets A and B. Indeed, we have $1 = A/B$, or alternatively $1 = A/(A \cap B)$, which is the set A from which the common part between A and B has been subtracted. Similarly, $2 = B/A = B/(A \cap B)$. Finally, we have: $3 = A \cap B$.

It is also useful to master the three set situation that is shown in Figure 2. There, $A = 1 \cup 4 \cup 5 \cup 7$, $B = 2 \cup 4 \cup 5 \cup 6$, and $C = 3 \cup 4 \cup 6 \cup 7$. Conversely, we see that $1 = A/(B \cup C)$, $2 = B/(A \cup C)$, and $3 = C/(A \cup B)$. Three surfaces of another kind are $5 = (A \cap B)/C$, $6 = (B \cap C)/A$, and $7 = (A \cap C)/B$. Finally,

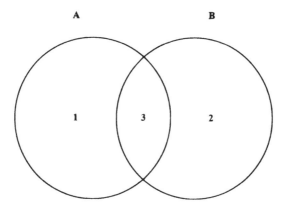

Figure 1 – Two Sets

the small surface in the center of the figure is the intersection of the three main sets: $4 = A \cap B \cap C$.

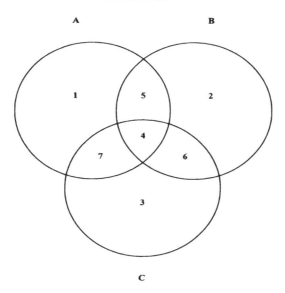

Figure 2 – Three Sets

While the two and three set situations should be known by

heart, the four set situation is less important. View it as an exercise and try to re-obtain the following results by yourself. Using the notation of Figure 3, we check that $A = 1 \cup 5 \cup 6 \cup 7 \cup 10 \cup 12 \cup 13 \cup 15$, $B = 2 \cup 5 \cup 8 \cup 9 \cup 10 \cup 12 \cup 14 \cup 15$, $C = 3 \cup 6 \cup 8 \cup 10 \cup 11 \cup 13 \cup 14 \cup 15$, and $D = 4 \cup 7 \cup 9 \cup 11 \cup 12 \cup 13 \cup 14 \cup 15$.

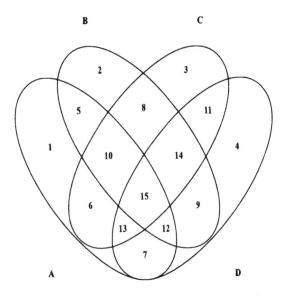

Figure 3 – Four Sets

Conversely, $1 = A/(B \cup C \cup D)$, $2 = B/(A \cup C \cup D)$, $3 = C/(A \cup B \cup D)$, and $4 = D/(A \cup B \cup C)$. Further, $5 = (A \cap B)/(C \cup D)$, $8 = (B \cap C)/(A \cup D)$, $11 = (C \cap D)/(A \cup B)$, $7 = (A \cap D)/(B \cup C)$, $6 = (A \cap C)/(B \cup D)$, and $9 = (B \cap D)/(A \cup C)$. We also have surfaces of the following kind: $10 = (A \cap B \cap C)/D$, $14 = (B \cap C \cap D)/A$, $12 = (A \cap B \cap D)/C$, and $13 = (A \cap C \cap D)/B$. Finally, the core intersection is $15 = A \cap B \cap C \cap D$.

Let now $\overline{A}$ be the **complementary set** of A within the bigger set Ω: $\overline{A} = \Omega/A$ or $\overline{A} = \Omega - A$. Equivalently, we have: $A \cup \overline{A} = \Omega$. The first important rules to remember about sets are the **De Morgan laws**:

$$\overline{A \cup B} = \overline{A} \cap \overline{B} \tag{5}$$

and
$$\overline{A \cap B} = \overline{A} \cup \overline{B}. \tag{6}$$

These laws are written as follows in the case of three sets:
$$\overline{A} \cup \overline{B} \cup \overline{C} = \overline{A \cap B \cap C} \tag{7}$$

and
$$\overline{A} \cap \overline{B} \cap \overline{C} = \overline{A \cup B \cup C}. \tag{8}$$

We can further generalize these laws to the case of an infinite number of sets A_i, for $i = 1, ..., +\infty$:
$$\bigcup_i \overline{A_i} = \overline{\bigcap_i A_i} \tag{9}$$

and
$$\bigcap_i \overline{A_i} = \overline{\bigcup_i A_i}. \tag{10}$$

Other important rules for sets are the **distributivity rules**:
$$(A \cup B) \cap C = (A \cap C) \cup (B \cap C) \tag{11}$$

and
$$(A \cap B) \cup C = (A \cup C) \cap (B \cup C). \tag{12}$$

These rules can be generalized as follows to the case of an infinite number of sets:
$$\left(\bigcup_i A_i \right) \cap C = \bigcup_i (A_i \cap C) \tag{13}$$

and
$$\left(\bigcap_i A_i \right) \cup C = \bigcap_i (A_i \cup C). \tag{14}$$

Now that we know the main rules for manipulating sets, we can introduce a measure for these sets. The measure P that we choose is of a specific type: it satisfies the **total probability rule** that states $P(\Omega) = 1$. The measure P is called a probability measure.

Recalling that $A \cup \overline{A} = \Omega$, the first property that we learn about P is the following:

$$P(A) + P(\overline{A}) = 1. \tag{15}$$

This property is used in numerous exercises, as can be seen from the tables in the Appendix. The following extension is also often useful:

$$P(A \cap B) + P(A \cap \overline{B}) = P(A). \tag{16}$$

Setting $A = \bigcap_i B_i$ and using Equations (9) and (15), we readily deduce the generalization of Equation (15) to the case of an infinite sequence of sets:

$$P\left(\bigcap_i B_i\right) + P\left(\bigcup_i \overline{B_i}\right) = 1. \tag{17}$$

Similarly, from Equations (10) and (15), we obtain another interesting generalization:

$$P\left(\bigcup_i B_i\right) + P\left(\bigcap_i \overline{B_i}\right) = 1. \tag{18}$$

The following result, which is probably the most important result of this section, avoids double counting when measuring unions of events:

$$P(A \cup B) = P(A) + P(B) - P(A \cap B). \tag{19}$$

This result is justified by the fact that $P(A) + P(B)$ counts $P(A \cap B)$ twice. See Figure 1.

The previous result can be generalized to the case of three sets:

$$\begin{aligned} P(A \cup B \cup C) = {}& P(A) + P(B) + P(C) \\ & - P(A \cap B) - P(A \cap C) - P(B \cap C) \\ & + P(A \cap B \cap C), \end{aligned} \tag{20}$$

where multiple counting is avoided in a similar way. See Figure 2.

Finally, let us denote by $A \ @ \ B$ the **disjointed union** of A and B. This is the union of A and B from which the joint set $A \cap B$ is taken out of. We have the important result:

$$P(A \ @ \ B) = P(A \cap \overline{B}) + P(\overline{A} \cap B). \tag{21}$$

2.2 Mutually exclusive events

We now come to a specific class of sets that has interesting properties. Take two sets A and B, also called events. These events are **mutually exclusive** when

$$A \cap B = \emptyset.$$

In that case, the probability measure of the intersection of A and B is null:

$$P(A \cap B) = 0. \tag{22}$$

Therefore, we have thanks to Equation (19):

$$P(A \cup B) = P(A) + P(B). \tag{23}$$

A countable collection of events $\{B_i\}$ is said to be mutually exclusive when we have:

$$\forall j \neq k \qquad B_j \cap B_k = \emptyset.$$

In that case, Equation (23) is generalized as follows:

$$P\left(\bigcup_i B_i\right) = \sum_i P(B_i). \tag{24}$$

If in addition we assume that the countable collection of mutually exclusive events $\{B_i\}$ is such that

$$\bigcup_i B_i = \Omega,$$

then we have
$$\sum_i P(B_i) = P(\Omega) = 1. \qquad (25)$$

A collection of events $\{B_i\}$ that are mutually exclusive and whose union is Ω is called a **partition** of Ω.

2.3 Addition and multiplication rules

We now examine the situation of events, or sets, that are **finite**. Let A be such an event. Then, Card(A) is the number of elements in A. "Card" stands for '**cardinality**', which essentially means 'size'.

When A and B are two finite events, we can define an **addition rule** as follows:

$$\text{Card}(A \cup B) = \text{Card}(A) + \text{Card}(B) - \text{Card}(A \cap B), \qquad (26)$$

which is analogous to Equation (19). When A and B are also mutually exclusive, this addition rule becomes

$$\text{Card}(A \cup B) = \text{Card}(A) + \text{Card}(B), \qquad (27)$$

which is analogous to Equation (23).

The quantity Card$(A \cup B)$ can be interpreted as an aggregate number of choices. The mutual exclusivity assumption means that there are Card(A) possible choices from A that are not related to the Card(B) possible choices from B. Therefore, there are indeed Card(A) + Card(B) choices from A *or* B.

Now, let the set $A \times B$ represent the collection of pairs (a, b), where a belongs to A and b belongs to B. This set is called the **Cartesian product** of A and B. We can now define a **multiplication rule** as follows:

$$\text{Card}(A \times B) = \text{Card}(A)\,\text{Card}(B). \qquad (28)$$

We interpret Card$(A \times B)$ as the number of ways of making both a choice in A *and* a choice in B. Now assume that A and B are both subsets of a universe Ω that is also finite. Then, the probability p of making both a choice in A *and* a choice in B can be expressed as follows:

$$p = \frac{\text{Card}(A \times B)}{\text{Card}(\Omega)} = \frac{\text{Card}(A)\,\text{Card}(B)}{\text{Card}(\Omega)}. \tag{29}$$

2.4 Independence of events

Events A and B are said to be independent, which we denote as $A \perp B$ when

$$P(A \cap B) = P(A)\,P(B). \tag{30}$$

First of all, note that if A and B are independent, then A and $\overline{B}$ are also independent [1]:

$$A \perp B \;\;\Rightarrow\;\; P(A \cap \overline{B}) = P(A)\,P(\overline{B}) \Leftrightarrow A \perp \overline{B}. \tag{31}$$

More generally, consider a family of events $\{A_i\}_{i=1,\cdots,n}$. These events are **two-by-two independent** when the property in Equation (30) holds for any pair of such events. Equivalently, this situation is described by

$$P(A_h \cap A_k) = P(A_h)\,P(A_k) \quad \forall h \neq k. \tag{32}$$

A more powerful property is the **mutual independence** of a collection of events $\{A_i\}_{i=1,\cdots,n}$. This property occurs when

$$P\left(\bigcap_{i=1}^{n} A_i\right) = \prod_{i=1}^{n} P(A_i). \tag{33}$$

Note that two-by-two independence (Equation (32)) might hold when mutual independence (Equation (33)) does not prevail. This is a classic source of exam questions.

---

1. From Equation (16), we have:

$$P(A \cap \overline{B}) = P(A) - P(A \cap B) = P(A) - P(A)P(B),$$

so that

$$P(A \cap \overline{B}) = P(A)(1 - P(B)) = P(A)P(\overline{B}).$$

Also note that

$$A \perp B \quad \Rightarrow \quad P(A \cup B) = P(A) + P(B) - P(A)\,P(B), \quad (34)$$

where this result should not be confounded with that of Equation (23).

Finally, assuming the mutual independence of events A, B, and C, we see that Equation (20) becomes

$$\begin{aligned}P(A \cup B \cup C) = {} & P(A) + P(B) + P(C) \\ & - P(A)\,P(B) - P(A)\,P(C) - P(B)\,P(C) \\ & + P(A)\,P(B)\,P(C). \end{aligned} \quad (35)$$

2.5 Combinatorial probability

Let $\mathcal{P}_n$ be the number of **permutations** of a set of n elements. The first important result in combinatorial probability is the following:

$$\mathcal{P}_n = n!. \quad (36)$$

Consider for instance the set $\{1, 2, 3\}$ that is comprised of $n = 3$ elements. We can construct $3! = 6$ permutations from this set: $(2, 1, 3)$, $(1, 3, 2)$, $(3, 2, 1)$, $(3, 1, 2)$, $(2, 3, 1)$, and $(1, 2, 3)$.

Next, we define the number of **unordered subsets**, or **combinations**, of k elements in a set of n elements:

$$\binom{n}{k} = \frac{n!}{k!\,(n-k)!}. \quad (37)$$

Consider again the set $\{1, 2, 3\}$ that is comprised of $n = 3$ elements. We can construct $\binom{3}{2} = \frac{3!}{2!\,1!} = 3$ combinations of two elements from this set: $(1, 2)$, $(1, 3)$, and $(2, 3)$.

The number of **ordered subsets**, or **k-permutations**, of k elements in a set of n elements, is used less. For clarity though, it is important to know this number, which is equal to $\frac{n!}{(n-k)!}$, so to $k!$ times the number of combinations.

In the case of the set $\{1, 2, 3\}$ that is comprised of $n = 3$ elements, we can construct $\frac{3!}{1!} = 6$ ordered subsets of $k = 2$

elements: $(1,2), (2,1), (1,3), (3,1), (2,3)$, and $(3,2)$. Thus, there are $2! = 2$ times more ordered subsets than unordered subsets in this case.

Coming back to unordered subsets, or combinations, it is important to know the following shortcuts:

$$\binom{n}{n} = \binom{n}{0} = 1 \qquad (38)$$

and

$$\binom{n}{n-1} = \binom{n}{1} = n. \qquad (39)$$

Finally, note that the sum of all the possible combinations has a simple expression:

$$\sum_{k=0}^{n} \binom{n}{k} = 2^n. \qquad (40)$$

This last property can be interpreted as follows. Assume you are drawing three random variables with values of zero or one. In total, there are $2^3 = 8$ possible outcomes: $(0,0,0)$, $(1,0,0)$, $(0,1,0)$, $(0,0,1)$, $(1,1,0)$, $(0,1,1)$, $(1,0,1)$, and $(1,1,1)$. Another classic example is that of a fruit salad. Assume you are given three fruits: apples, pears, and oranges. How many types of fruit salads can you prepare? Again, the answer is eight and takes into account the fact that you may or may not include each fruit in the salad.

2.6 Conditional probability

Let us now introduce the simplest and most fundamental form of **conditional probability** that is examined in this book. Let A and B be two events. Then, the probability that A occurs when you know that B has occurred is denoted by $P(A|B)$. This probability satisfies

$$P(A|B) = \frac{P(A \cap B)}{P(B)} \quad \Leftrightarrow \quad P(A \cap B) = P(A|B)\, P(B), \qquad (41)$$

where, in practice, the two equivalent forms above are equally useful.

An interesting subcase of this formula is

$$A \subset B \quad \Rightarrow \quad P(A|B) = \frac{P(A)}{P(B)}, \tag{42}$$

which follows from $P(A \cap B) = P(A)$ in this case.

These two formulas can be illustrated as follows. Assume that the probability of drawing a circle is one half and that the probability of drawing a square is one half. Next, assume that the probability that the plot is blue is one half and that the probability that the plot is red is one half. These two random draws are further assumed to be independent.

What is the probability of drawing a red plot (event A) when we know that we drew a circle (event B)? From Equation (41), the answer is $P(A \cap B)/P(B)$. By independence, we have $P(A \cap B) = P(A) \, P(B)$. Therefore, the solution is $(P(A)P(B))/P(B) = P(A) = 1/2$.

Now, what is the probability of drawing a red circle (event C) when we know that we drew a circle (event B)? Red circles represent one fourth of all the possible plots, so $P(C) = \frac{1}{4}$. The probability of drawing a circle is $P(B) = \frac{1}{2}$. C is indeed a subevent of B. Therefore, the result follows from Equation (42) and is $P(C|B) = (1/4)/(1/2) = 1/2$.

Dividing Equation (16) by $P(A)$ yields a new formulation of this equation:

$$1 = P(B|A) + P(\overline{B}|A). \tag{43}$$

Finally, you should spend some time examining the following extension of Equation (41) that is sometimes useful for the most complex exercises:

$$P(A|B \cap C) = \frac{P(A \cap B|C)}{P(B|C)}. \tag{44}$$

2.7 The law of total probability and Bayes' theorem

We are first interested in generalizing Equation (16) to the case of a collection of events. This generalization is provided by the law of total probability that can be expressed in two forms. The first form is as follows:

Law of total probability (first form)

Let $\{B_i\}$ be a partition of Ω, that is, a countable collection of mutually exclusive events such that $\bigcup_i B_i = \Omega$. For any event A,

$$P(A) = \sum_i P(A \cap B_i). \tag{45}$$

Alternatively, using the definition (41) of conditional probabilities, we can write:

Law of total probability (second form)

Let $\{B_i\}$ be a partition of Ω. For any event A,

$$P(A) = \sum_i P(A|B_i) P(B_i). \tag{46}$$

A huge part of modern science (statistical learning, decision theory, some non-life insurance theories...) relies on the Bayes theorem, which is a tool to reverse conditional probabilities. Namely, this theorem allows us to express $P(B|A)$ as a function of $P(A|B)$. This is achieved as follows:

Bayes' theorem (first form)

For any events A and B,

$$P(B|A) = \frac{P(A|B) P(B)}{P(A)}. \tag{47}$$

This result is a direct consequence of Equation (41), which allows us to write $P(A \cap B) = P(A|B)P(B) = P(B|A)P(A)$. This theorem admits a more detailed form that is the key tool for solving a variety of exercises:

Bayes' theorem (second form)

Let $\{B_i\}$ be a partition of Ω. For any events A and B,

$$P(B|A) = \frac{P(A|B)P(B)}{\sum_i P(A|B_i)P(B_i)}. \qquad (48)$$

When you prepare for the C/STAM exam, you will use this theorem extensively.

Chapter 3

Univariate probability distributions

This chapter introduces probability mass functions, probability density functions, and cumulative distribution functions. It defines and studies moments, means, variances, modes, percentiles, and moment generating functions. This chapter is constructed around the main random variables: binomial, negative binomial, geometric, hypergeometric, Poisson, uniform, exponential, gamma, normal, and mixed. The transformations of random variables are also examined.

3.1 Probability mass and density functions

This section covers the study of random variables in the univariate case (in dimension 1). A clean mathematical definition of random variables is out of the scope of this book. Just understand that a random variable X is a function that associates a quantity $X(\omega)$ with a state of the world ω.

In fact, we will define random variables via their probability mass function or their probability density function. Or we will often say that we characterize random variables by their **probability distributions**. Two main situations will be encountered: the discrete case, where random variables take values in a discrete set - $\mathbb{N}$ for instance - and the continuous case, where random variables takes values in a continuous set - $\mathbb{R}$ for instance. On rare occasions, we will be dealing with mixed situations.

In the discrete setting, a **probability mass function** g is defined as follows:
$$g(x) = P(X = x).$$

In the continuous setting, a **probability density function** f is defined as follows:
$$f(x)\,dx = P(x \leq X < x + dx),$$

where dx is an infinitesimal increment around x.

It is meaningless to use the probability mass function in the continuous setting because the probability of reaching a value exactly - for instance 2.56723 - is null and irrelevant in this case.

Instead, the approach of probability density functions allows us to measure the probability associated with any interval I in a simple way:
$$P(X \in I) = \int_{x \in I} f(x)\,dx,$$

where this probability is computed by using the standard integration theory.

We can also define a **conditional density function** as follows:
$$f(x|A) = \frac{f(x)\,\mathbb{1}_{x \in A}}{P(A)}, \qquad (49)$$

where $\mathbb{1}_{x \in A}$ is the indicator function that equals one if $x \in A$ and zero if $x \notin A$. This formula provides a generalization of Equation (41). Equivalently, we will also often write
$$f(x|A) = \frac{f(x)}{P(A)}, \quad \forall x \in A. \qquad (50)$$

We now come to the definitions of the main random variables that are used in actuarial and financial applications. Here we note that the random variables X and Y are said to be **independent** when $P(X \leq x \cap Y \leq y) = P(X \leq x)\,P(Y \leq y)$ for all x and y.

Discrete random variables

The simplest possible random variable takes only two values: one with probability p and zero with probability $1-p$.

This random variable is called a **Bernoulli** random variable and is said to follow a Bernoulli probability distribution with parameter p. It is classically denoted by $N \sim \mathcal{B}(p)$, where the $\sim$ operator indicates that the random variable N is associated with the probability distribution $\mathcal{B}(p)$.

It is possible to define a **generalized Bernoulli** random variable that takes the following values: a with probability p and b with probability $1-p$. This random variable is in turn denoted by $N \sim \mathcal{B}_{a,b}(p)$.

The next type of random variable that we study is the **binomial** random variable $N \sim \mathcal{B}(n,p)$. Its probability distribution depends on two parameters: n and p. In fact, **a binomial random variable is the sum of** n **independent Bernoulli random variables** with parameter p. The binomial probability distribution, or probability mass function, is as follows:

$$P(N = k) = \binom{n}{k} p^k q^{n-k}, \qquad (51)$$

where $q = 1 - p$. This quantity describes for instance the probability of a **coin** landing k times on heads, out of n draws.

Although it is not an individual random variable, we can mention the **multinomial** random variable $\mathbf{N} \sim \mathcal{B}(n, \{p_i\}_{i=1,\cdots,m})$, which generalizes the binomial random variable. The multinomial probability distribution is given by

$$P(N_1 = k_1 \cap \cdots \cap N_m = k_m) = \frac{n!}{k_1! \cdots k_m!} p_1^{k_1} \cdots p_m^{k_m}, \qquad (52)$$

with $\sum_{i=1}^{m} k_i = n$, which generalizes the binomial probability distribution given in Equation (51). When $m = 6$ for instance, this quantity measures the probability that a **die** shows k_1 times a one, k_2 times a two, and so on, out of n draws.

Another important probability distribution is the **geometric** distribution. Assuming a sequence of Bernoulli draws is launched, a geometric random variable counts the number of trials that are needed to achieve a first **success** (to draw a one). We include in this number of trials the trial corresponding to the first success. We denote by $N \sim \mathcal{G}(p)$ such a random variable that is characterized by

$$P(N = k) = q^{k-1}p. \tag{53}$$

In this expression, we measure the probability that the number of trials N is k. In this scenario, the first success is the kth trial and has probability p. This success requires $k - 1$ failures beforehand, each with a probability $q = 1 - p$, which explains the form of the probability mass function.

A related distribution is the **negative binomial** distribution. In its **first form**, it counts the number N of **failures** (zeros) that are observed until r successes (ones) are recorded. Denoting by $N \sim \mathcal{NB}_1(r, p)$ a negative binomial random variable of the first type, we have:

$$P(N = h) = \binom{h + r - 1}{h} p^r q^h. \tag{54}$$

There is a **second form** of the **negative binomial** distribution. In this second form, we count the number of **trials** N' that are required until r successes have been reached. We include in this random variable the last success, as in the geometric distribution case. We introduce the notation $N' \sim \mathcal{NB}_2(r, p)$ and we write:

$$P(N' = n) = \binom{n - 1}{r - 1} p^r q^{n-r}. \tag{55}$$

Note that the negative binomial distribution boils down to the geometric distribution in the case of $r = 1$.

An important probability distribution in insurance theory is the **Poisson** distribution. A Poisson random variable, which we denote by $N \sim \mathcal{P}(\lambda)$, has the following probability mass function:

$$P(N = k) = \frac{e^{-\lambda}\lambda^k}{k!}. \tag{56}$$

A Poisson random variable is typically used to represent a number of claims. The Poisson, binomial, and negative binomial distributions constitute the three main **counting distributions** that are used for this purpose. You will work a lot with these distributions when you prepare for Exam STAM.

We now construct the **hypergeometric** distribution. Consider a collection of $a + b$ elements pertaining to two categories. There are a elements in the first category and b elements in the second category. We draw n elements from the collection and we want to compute the probability that k of these elements pertain to the first category. This probability is equal to

$$P(N = k) = \frac{\binom{a}{k}\binom{b}{n-k}}{\binom{a+b}{n}}, \qquad (57)$$

where $N \sim \mathcal{H}(a, b, n)$ denotes a hypergeometric random variable.

Similar to the binomial/multinomial case, it is also possible to construct a **multivariate hypergeometric** distribution that extends the hypergeometric distribution. Consider n categories with a_i elements in each category $i = 1, \cdots, n$ and A elements in total, so that $A = \sum_{i=1}^{n} a_i$. We draw K elements in total and we want to compute the probability to draw k_i elements in each category $i = 1, \cdots, n$. We have:

$$P(N_1 = k_1, \cdots, N_n = k_n) = \frac{\prod_{i=1}^{n} \binom{a_i}{k_i}}{\binom{A}{K}}, \qquad (58)$$

where, for each i, N_i is the number of elements drawn from category i and where $K = \sum_{i=1}^{n} k_i$.

Continuous random variables

The simplest continuous random variable is the **uniform** random variable, which we denote by $U \sim \mathcal{U}(a, b)$. The density function associated with its distribution is constant over a bounded

interval:
$$f_U(x) = \frac{1}{b-a}, \qquad x \in [a,b]. \tag{59}$$

An interesting subcase is the **standard uniform** random variable $V \sim \mathcal{U}(0,1)$ defined by the density:
$$f_V(x) = 1, \qquad x \in [0,1]. \tag{60}$$

Note that most software can generate this random variables (by calling 'rnd', 'rand', ..., routines). Moreover, most continuous random variables can be simulated as functions of standard uniform random variables.

A probability distribution that you will use extensively in Exam LTAM is the **exponential** distribution. We denote an exponential random variable as $X \sim \mathcal{E}(\theta)$ and we associate it with the following density:
$$f_X(x) = \frac{1}{\theta}\, e^{-\frac{x}{\theta}}, \qquad x \geq 0. \tag{61}$$

Be careful that a different convention exists where $\theta := \frac{1}{\theta}$. It is preferable to use the convention of Equation (61) because it is widespread, it is used in the tables of the subsequent Exam STAM, and it makes more sense (as we will see later on, θ is identical to the mean of the distribution in this convention).

The exponential distribution is **memoryless**:
$$X - d \mid X \geq d \stackrel{\text{law}}{=} X, \tag{62}$$

where this equality states that the excess of X beyond a threshold d, when you know that X exceeds this threshold, has the same distribution (or law) as X.

An extension of the exponential distribution is the **gamma** distribution. We denote as $X \sim \mathcal{G}(\alpha, \theta)$ a gamma random variable that is defined by the density:
$$f_X(x) = \frac{\left(\frac{x}{\theta}\right)^{\alpha} e^{-\frac{x}{\theta}}}{x\, \Gamma(\alpha)}, \qquad x \geq 0, \tag{63}$$

where Γ is the gamma function.

When α is an integer, a gamma random variable is simply the sum of α independent exponential random variables. In that specific case, the gamma distribution is called an **Erlang** distribution. When $\alpha = 1$, you can check that the density in (63) simplifies to the density in (61).

The last continuous distribution that you will use is the **normal** - or **Gaussian** - distribution. A normal random variable denoted by $X \sim \mathcal{N}(\mu, \sigma)$ is associated with the following density function:

$$f_X(x) = \frac{1}{\sigma\sqrt{2\pi}} e^{-\frac{(x-\mu)^2}{2\sigma^2}}, \qquad x \in \mathbb{R}. \qquad (64)$$

Normal random variables are used in all fields of science, from physics to botany and from finance to geology. Although the normal distribution is increasingly challenged, it is extremely important to be able to perform the classic computations associated with this distribution.

Mixed random variables

Mixed random variables have both a discrete and a continuous component. Their probability function f is the sum of a probability mass component f_d and a probability density component f_c. In this construction, we take care that the total probability, which is the sum of f_d plus the integral of f_c over their domain of definition, remains equal to one. A simple example of such a random variable is defined by the density:

$$f(x) = f_d(x) + f_c(x) = \frac{1}{2}\mathbb{1}_{x=2} + \frac{1}{2}\mathbb{1}_{x\in[0,1]},$$

where the random variable takes the value two with probability one half, and its takes a value that is uniformly distributed between zero and one again with probability one half.

Note carefully that constructing a density that is the sum of two functions is different from summing up two random variables associated with these functions. So, the mixed random variable that we consider here is *not* the sum of a random variable that has f_d as its probability mass function and of another random variable that has f_c as its probability density function. By the way, each of these two functions do not sum up to one over their domain of definition in full generality, so they cannot be called a probability mass function or a probability density function in the strict sense.

3.2 Cumulative distribution functions

A **cumulative distribution function**, or **cdf**, sums up or integrates the values of a probability mass or density function *below* a given threshold. In the case of a discrete random variable N, this amounts to computing the function F_N such as:

$$F_N(n) = P(N \leq n) = \sum_{k=k_{\min}}^{n} P(N = k), \qquad (65)$$

where $k_{\min}$ is the smallest value that the random variable N can take. In most applications $k_{\min} = 0$.

The total probability, which is classically equal to one, can be recovered as follows:

$$F_N(k_{\max}) = \sum_{k=k_{\min}}^{k_{\max}} P(N = k) = 1, \qquad (66)$$

where $k_{\max}$ is the largest value that the random variable N can take. In most applications $k_{\max} = +\infty$.

Individual probabilities can be easily recovered from cumulative distribution functions:

$$P(N = n) = P(N \leq n) - P(N \leq n-1) = F(n) - F(n-1). \quad (67)$$

In the case of a continuous random variable X, the cumulative distribution function is a function F_X that satisfies:

$$F_X(x) = P(X \leq x) = \int_a^x f_X(s)\, ds, \qquad (68)$$

where a is the smallest value attainable by X. In most applications $a = 0$ or $a = -\infty$.

The total probability is now obtained as follows:

$$F_X(b) = \int_a^b f_X(s)\,ds = 1, \tag{69}$$

where b is the largest value attainable by X. In most applications $b = 1$ or $b = +\infty$.

We can recover the probability density function by differentiating the cumulative distribution function:

$$f_X(s) = \left(\frac{dF_X(x)}{dx}\right)_{x=s}. \tag{70}$$

The probability that a random variable X has values in a range is the difference between two values of its cumulative distribution function:

$$P(\alpha < X \leq \beta) = P(X \leq \beta) - P(X \leq \alpha) = F(\beta) - F(\alpha). \tag{71}$$

A **survival function** sums up or integrates the values of a probability function *above* a given threshold. In the continuous case, we have:

$$S_X(x) = P(X > x) = \int_x^b f_X(t)\,dt. \tag{72}$$

The total probability is in turn recovered as follows:

$$S_X(a) = 1. \tag{73}$$

In the discrete case, the survival function becomes

$$S_N(n) = P(N > n) = \sum_{k=n+1}^{k_{\max}} P(N = k) = 1 - \sum_{k=k_{\min}}^{n} P(N = k). \tag{74}$$

Note that the cumulative distribution and survival functions can be readily recovered one from the other:

$$S_X(x) = 1 - F_X(x), \tag{75}$$

where the result holds whether the underlying random variable is continuous or discrete.

Classic examples in the discrete case

It is important to be able to quickly write the cumulative distribution function of a discrete random variable correctly. For instance, in the case of a binomial random variable $N \sim \mathcal{B}(n,p)$, the cumulative distribution function is

$$P(N \leq k) = \sum_{i=0}^{k} \binom{n}{i} p^i (1-p)^{n-i} \tag{76}$$

and the survival function is

$$P(N > k) = \sum_{i=k+1}^{n} \binom{n}{i} p^i (1-p)^{n-i}, \tag{77}$$

where n is the largest value attainable by the binomial random variable.

Similarly, when N follows a Poisson distribution $\mathcal{P}(\lambda)$, the cumulative distribution function is

$$P(N \leq k) = e^{-\lambda} \sum_{i=0}^{k} \frac{\lambda^i}{i!} \tag{78}$$

and the survival function is

$$P(N > k) = e^{-\lambda} \sum_{i=k+1}^{+\infty} \frac{\lambda^i}{i!}, \tag{79}$$

where the largest value attainable by the Poisson random variable is infinite.

In the case of a geometric random variable $N \sim \mathcal{G}(p)$, there are simple closed-form expressions that you should know by rote:

$$P(N \leq k) = 1 - q^k \tag{80}$$

and

$$P(N > k) = q^k. \tag{81}$$

Classic examples in the continuous case

In this part, we first consider the case of a uniform random variable $U \sim \mathcal{U}(a,b)$. Its cumulative distribution function has a particularly simple form:

$$P(U \leq u) = \frac{u-a}{b-a}, \qquad u \in [a,b], \tag{82}$$

where this cdf is null below a and equal to one above b.

For the standard uniform random variable $V \sim \mathcal{U}(0,1)$, the cdf is even simpler:

$$P(V \leq v) = v, \qquad x \in [0,1]. \tag{83}$$

One of the most important closed-form expressions to know is the cdf of an exponential random variable $X \sim \mathcal{E}(\theta)$:

$$F(x) = 1 - e^{-\frac{x}{\theta}}, \qquad x \geq 0. \tag{84}$$

This cumulative distribution function is the key to solving a great number of exercises. The survival function of an exponential random variable follows accordingly:

$$S_X(x) = e^{-\frac{x}{\theta}}, \qquad x \geq 0. \tag{85}$$

The following result is extremely useful for the study of insurance contracts in the presence of a deductible and exponentially distributed claims:

$$(X-d)\mathbb{1}_{X \geq d} \stackrel{\text{law}}{=} (X-d \mid X \geq d)P(X \geq d) \stackrel{\text{law}}{=} Xe^{-\frac{d}{\theta}}, \tag{86}$$

where d stands for the deductible and where the first term in this series of equalities corresponds to what is paid by an insurer to an insured after the occurrence of a loss X that exceeds the deductible.

The first equality in (86) is analogous to Equation (41) where we interpret the indicator operator $\mathbb{1}$ as an intersection operator $\cap$. While there will be more details about this analogy in the next chapter, you should make note of it now because it is very useful. The second equality in (86) follows from Equations (62) and (85).

You should also note that

$$(X - d)\mathbb{1}_{X \geq d} = \max(X - d, 0) = (X - d)^+.$$

The probability that an exponential random variable belongs to a range has a simple form:

$$P(\alpha < X \leq \beta) = F(\beta) - F(\alpha) = e^{-\frac{\alpha}{\theta}} - e^{-\frac{\beta}{\theta}} \quad \forall \beta \geq \alpha \geq 0, \tag{87}$$

where this result follows from Equations (71) and (84).

Let us now come to the cumulative distribution functions of normal random variables. Set $X \sim \mathcal{N}(\mu, \sigma)$ and $Z \sim \mathcal{N}(0,1)$. The following series of equalities is useful for Exam P/1 and for most of the subsequent exams; it should absolutely be mastered:

$$P(X \leq c) = P\left(\frac{X - \mu}{\sigma} \leq \frac{c - \mu}{\sigma}\right) = P\left(Z \leq \frac{c - \mu}{\sigma}\right) = N\left(\frac{c - \mu}{\sigma}\right), \tag{88}$$

where the function N is by convention the cdf of a standard normal random variable $Z \sim \mathcal{N}(0,1)$ and where we use the fact that $\frac{X-\mu}{\sigma}$ has the same distribution as Z.

Similar equalities can be derived for survival functions:

$$P(X > c) = 1 - P(X \leq c) = 1 - N\left(\frac{c - \mu}{\sigma}\right) = N\left(-\frac{c - \mu}{\sigma}\right), \tag{89}$$

where we use the important property:

$$N(-x) = 1 - N(x). \tag{90}$$

The probability that a normal random variable $X \sim \mathcal{N}(\mu, \sigma)$ belongs to a range $(\alpha, \beta]$ is given by

$$P(\alpha < X \leq \beta) = P\left(\frac{\alpha - \mu}{\sigma} < Z \leq \frac{\beta - \mu}{\sigma}\right) = N\left(\frac{\beta - \mu}{\sigma}\right) - N\left(\frac{\alpha - \mu}{\sigma}\right). \tag{91}$$

You should also note that for a standard normal random variable $Z \sim \mathcal{N}(0,1)$, the following result holds:

$$P(|Z| \leq c) = 2P(Z \leq c) - 1 = 2N(c) - 1. \tag{92}$$

Mixed random variables

Recall that mixed random variables have both continuous and discrete components. Their cumulative distribution function has the following form:

$$F(x) = \int_a^x f_c(s)\, ds + \sum_{k=k_{\min}}^{n(x)} P(X = x_k), \qquad (93)$$

where $n(x)$ is the largest integer such that the x_k's in the sum are smaller than or equal to x. The latter condition can be written as follows:

$$n(x) = \max(k \mid x_k \leq x).$$

Setting $x = b$, we recover the total probability as follows:

$$F(b) = \int_a^b f(s)\, ds + \sum_{k=k_{\min}}^{k_{\max}} P(X = x_k) = 1. \qquad (94)$$

3.3 Mode, percentiles, median, mean, and moments

A probability distribution is characterized by its mode, percentiles, median, mean, and its moments. These indicators provide useful measures of risk when probability distributions are used in insurance or finance. Conditional expectations are also examined in this section.

Mode

The **mode** of a probability distribution is the value taken with maximum probability by its underlying random variable. For a continuous distribution, it is the value x^* that maximizes the density (when this latter quantity exists):

$$f'(x^*) = 0, \qquad f''(x^*) \leq 0. \qquad (95)$$

Percentiles

The pth **percentile** of a continuous probability distribution is the value x^p that solves:

$$F(x^p) = \int_a^{x^p} f_X(x)\, dx = \frac{p}{100}. \tag{96}$$

In the discrete case, it might not be possible to find an exact value x^p such that the cdf taken at this value equals $\frac{p}{100}$. Instead, we can define the pth percentile of a discrete probability distribution as the smallest value x^p that satisfies

$$P(X \leq x^p) \geq \frac{p}{100}. \tag{97}$$

The **median** of a probability distribution is just its 50th percentile.

One can usefully and readily check that the pth percentile $x_{\mathcal{U}}^p$ of $U \sim \mathcal{U}(a,b)$ satisfies

$$x_{\mathcal{U}}^p = a + p\,(b-a). \tag{98}$$

Then, the pth percentile $x_{\mathcal{N}}^p$ of $\mathcal{N}(\mu, \sigma)$ satisfies

$$x_{\mathcal{N}}^p = \mu + z^p\, \sigma, \tag{99}$$

where z^p is the pth percentile of $\mathcal{N}(0,1)$.

Note that $z^p = N^{-1}(p/100)$, where N is the standard normal cdf. When $p \geq 50$, z^p can be found in the table of standard normal cdfs provided by the SOA. The number $p/100$ is read in both the row and the column of this table. When $p < 50$, you first need to compute z^{100-p} and then you set $z^p = -z^{100-p}$. This result follows from Equation (90).

Table 1 gives you the main standard normal percentiles that are used in actuarial and financial applications. These should be known by heart.

Now assume that you are given a value x that is drawn from a random variable $X \sim \mathcal{N}(\mu, \sigma)$ and you want to know to which percentile this value corresponds to. First, you can compute the **z-score**:

$$z - \text{score} = \frac{x - \mu}{\sigma}, \tag{100}$$

p	90	95	99
z^p	1.282	1.645	2.326

Table 1 – Main Standard Normal Percentiles

and then you simply need to read this number in the table of standard normal cdfs provided in the exam to obtain $p/100$ from the row and column of the table.

You may want to **interpolate** to obtain a more precise value of p. This can be achieved as follows: Assume that you are given $z^p \in (z^{p_1}, z^{p_2})$, where z^{p_1} is the rounded value of z^p to the next smaller percent and z^{p_2} is the rounded value of z^p to the next larger percent. Further, z^{p_1} is the p_1^{th} percentile of $\mathcal{N}(0,1)$ and z^{p_2} is the p_2^{th} percentile of $\mathcal{N}(0,1)$. Then, an approximation of p is given by

$$p = p_1 + \frac{z^p - z^{p_1}}{z^{p_2} - z^{p_1}} (p_2 - p_1). \tag{101}$$

Mean

A **mean**, or **expectation**, of a random variable is an arithmetic average of the values taken by the random variable, where these values are weighted by their probabilities. We first define the expectation of a discrete random variable:

$$E(N) = \sum_{k=k_{\min}}^{k_{\max}} k \, P(N=k). \tag{102}$$

By analogy, the expectation of a continuous random variable is

$$E(X) = \int_a^b x \, f(x) \, dx, \tag{103}$$

where the weight of x is $f(x) \, dx$ that is the probability associated with an infinitesimal interval around x.

An insurer makes a payment of $\max(X - d, 0)$ to an insured when a claim of size X occurs and is subject to a deductible d.

We can compute the expected claim payment in the presence of a deductible d in the continuous case:

$$E(\max(X - d, 0)) = \int_d^b (x - d) \, f(x) \, dx. \tag{104}$$

But, in the discrete case, we have:

$$E(\max(X - d, 0)) = \sum_i (x_i - d)^+ \, P(X = x_i). \tag{105}$$

When a claim of size X occurs and is subject to a deductible d, the amount $\min(X, d)$ remains unreimbursed by the insured. In the continuous case, the expected unreimbursed loss in the presence of a deductible d is

$$E(\min(X, d)) = \int_a^b \min(x, d) \, f(x) \, dx = \int_a^d x \, f(x) \, dx + \int_d^b d \, f(x) \, dx \tag{106}$$

or

$$E(\min(X, d)) = \int_a^d x \, f(x) \, dx + d \, P(X > d). \tag{107}$$

Note that $E(\min(X, u))$, which is the formula for the expected claim payment by an insurer to an insured in the presence of a maximum reimbursement u and without a deductible, can also be computed using Equation (107). Remember this formula and be extremely careful not to forget the last term (a classic mistake).

Finally, the expected claim payment by an insurer to an insured in the presence of both a deductible d and a maximum reimbursement u can be written in the following form:

$$E\left(\min\left((X - d)^+, u\right)\right) = \int_d^{d+u} (x - d) \, f(x) \, dx + u \int_{d+u}^b f(x) \, dx, \tag{108}$$

assuming $u \in [a, b]$.

Means of discrete distributions

We now come to the known closed-form expressions of the means of discrete distributions. Let $N \sim \mathcal{B}(p)$ be a standard Bernoulli random variable. Its mean is simply equal to the probability of success (drawing a one):

$$E(N) = p. \tag{109}$$

Next, let $N \sim \mathcal{B}_{a,b}(p)$ be a generalized Bernoulli random variable that takes the following values: a with probability p and b with probability $1 - p$. Then, the mean becomes

$$E(N) = p\,a + b\,(1-p), \tag{110}$$

where this expression is a very useful formula for exams STAM and LTAM.

For a binomial random variable $N \sim \mathcal{B}(n,p)$, which is defined as the sum of n independent standard Bernoulli random variables, the mean is equal to n times the mean of a standard Bernoulli distribution:

$$E(N) = n\,p. \tag{111}$$

For a geometric random variable $N \sim \mathcal{G}(p)$, which describes the number of trials that are needed to achieve one success, we have:

$$E(N) = \frac{1}{p}, \tag{112}$$

where the smaller the probability of success, the longer it takes on average before a success is obtained.

For a coin, $p = 1/2$ and it takes on average two trials before a heads is drawn. For a die, $p = 1/6$ and it takes on average six trials before a one is drawn, and so on.

Let $N \sim \mathcal{NB}_1(r,p)$ be a negative binomial random variable of the first type that describes the number of failures until r successes have been reached. Its expectation is

$$E(N) = \frac{r\,q}{p}. \tag{113}$$

Note that for Exam STAM, the notation is changed to $p = \frac{1}{1+\beta}$ and $q = \frac{\beta}{1+\beta}$, and the mean becomes $E(N) = r\,\beta$, which is similar to the expectation of a binomial random variable given in Equation (111).

Let $N' \sim \mathcal{NB}_2(r, p)$ be a negative binomial random variable of the second type that now describes the number of trials until reaching r successes. We have:

$$E(N') = \frac{r}{p} \tag{114}$$

You can check that $E(N') = E(N) + r$. This is consistent with the fact that the number of trials is equal to the number of failures plus the number of successes ($N' = N + r$).

In the case of a Poisson random variable, the mean is simply equal to the parameter of the probability distribution:

$$E(X) = \lambda. \tag{115}$$

Note that the Poisson distribution $\mathcal{P}(\lambda)$ can be used to approximate the Binomial distribution $\mathcal{B}(n, p)$. This is achieved by making the means of the two distributions equal:

$$\lambda = n\,p. \tag{116}$$

This method is valid when n is sufficiently large and p is sufficiently small. We typically require $n \geq 20$ and $p \leq 0.05$.

The last mean of a discrete distribution that you must know is the mean of a hypergeometric random variable $N \sim \mathcal{H}(a, b, n)$. The expected number of draws in a given first category is equal to

$$E(N) = n\,\frac{a}{a+b}, \tag{117}$$

where we see that the expected number of draws in the first category is a proportion of the total number of draws n. The proportionality coefficient $\frac{a}{a+b}$ is simply the ratio of the number of elements from the first category to the total number of elements.

Means of continuous distributions

The mean of a uniform random variable $U \sim \mathcal{U}(a, b)$ is simply the midpoint of the domain of definition:

$$E(U) = \frac{a+b}{2}. \tag{118}$$

Similarly, in the subcase of a standard uniform random variable $V \sim \mathcal{U}(0, 1)$, we have:

$$E(V) = \frac{1}{2}. \tag{119}$$

For an exponential random variable, the mean is merely equal to the parameter:

$$E(X) = \theta. \tag{120}$$

Therefore, if you are given the mean of an exponential random variable, you readily know the value of its parameter. Also note that there is a simple link between the median and the mean of an exponential random variable:

$$\text{Median}(X) = E(X)\ \ln(2) = \theta\ \ln(2). \tag{121}$$

Using Equation (86), we can now fully compute the value of the expected claim payment in the presence of a deductible and when the underlying risk has an exponential distribution:

$$E(\max(X - d, 0)) = E((X-d)\mathbb{1}_{X \geq d}) = e^{-\frac{d}{\theta}} E(X) = \theta\ e^{-\frac{d}{\theta}}. \tag{122}$$

For a gamma random variable $X \sim \mathcal{G}(\alpha, \theta)$, the mean is equal to the product of the two parameters:

$$E(X) = \alpha\ \theta.$$

Finally, for a normal random variable $X \sim \mathcal{N}(\mu, \sigma)$, the mean is simply equal to the first parameter:

$$E(X) = \mu.$$

Mean of a mixed distribution

The mean of a mixed distribution can be computed as follows:

$$E(X) = \sum_i x_i \, P(X = x_i) + \int_a^b x \, f_X(x) \, dx, \qquad (123)$$

where we recall that the total probability is recovered from

$$\sum_i P(X = x_i) + \int_a^b f_X(x) \, dx = 1.$$

Conditional expectation

We often need to work with expectations that are conditional on an event or on the value of a random variable. In the continuous case, a **conditional expectation** is defined as follows:

$$E(X \mid A) = \int_a^b x \, f(x|A) \, dx. \qquad (124)$$

Using Equation (49), we can also write:

$$E(X \mid A) = \frac{\int_a^b x \, f(x) \, \mathbb{1}_A \, dx}{P(A)}. \qquad (125)$$

In the discrete case, the conditional expectation becomes

$$E(N \mid A) = \sum_{k=k_{\min}}^{k_{\max}} k \, P(N = k \mid A). \qquad (126)$$

A classic example of computation in the discrete case is shown below:

$$E(N \mid N \geq n) = \sum_{k=k_{\min}}^{k_{\max}} k \, P(N = k \mid k \geq n) = \frac{\sum_{k \geq n} k \, P(N = k)}{P(k \geq n)}, \qquad (127)$$

where the second equality is a consequence of Equation (41).

A key result is the **tower property** of conditional expectations:
$$E(X) = E(E(X|A)). \tag{128}$$

Let $\{B_i\}$ be a countable collection of mutually exclusive events. Then, we also have:
$$E(X) = \sum_i E(X|B_i) P(X \in B_i), \tag{129}$$

which is an extension of the law of total probability shown in (46).

Non-centered moments

The results for the mean can be readily extended to the case of non-centered moments. The **non-centered moment** of order h of a discrete random variable is defined as follows:
$$E(N^h) = \sum_{k=k_{\min}}^{k_{\max}} k^h \, P(N = k). \tag{130}$$

The non-centered moment of order h of a continuous random variable is in turn
$$E(X^h) = \int_a^b x^h \, f(x) \, dx. \tag{131}$$

In the continuous case, the moment of order h of the claim payment made by an insurer to an insured is in the presence of a deductible d:
$$E((\max(X - d, 0))^h) = \int_d^b (x - d)^h \, f(x) \, dx. \tag{132}$$

The moment of order h of the unreimbursed loss - amount not covered by the insurer - is in the presence of a deductible d:
$$E\left((\min(X, d))^h\right) = \int_a^d x^h \, f(x) \, dx + d^h \, P(X > d). \tag{133}$$

Note that the latter quantity is also equal to the moment of order h of the claim payment made by an insurer to an insured in the presence of a maximum reimbursement u, where we only need to replace d with u in Equation (133).

The moment of order h of the claim payment made by an insurer to an insured is in the presence of both a deductible d and a maximum reimbursement u:

$$E\left(\left(\min\left((X-d)^+, u\right)\right)^h\right) = \int_d^{d+u} (x-d)^h f(x)\,dx + u^h \int_{d+u}^b f(x)\,dx, \tag{134}$$

assuming $u \in [a, b]$.

We now give classic examples of non-centered moments that you will encounter when solving exercises. For a Bernoulli random variable $N \sim \mathcal{B}(p)$, we have:

$$E\left(N^2\right) = p \tag{135}$$

and, more generally [1] for any h:

$$E\left(N^h\right) = p. \tag{136}$$

For a Poisson random variable, remember that

$$E(X^2) = \lambda + \lambda^2. \tag{137}$$

The moments of exponential random variables are also quite simple:

$$E(X^2) = 2\,\theta^2 \tag{138}$$

and

$$E(X^h) = h!\,\theta^h. \tag{139}$$

Keeping the exponential random variable assumption for X, we obtain the following generalization of Equation (122):

$$E\left((X-d)^h \mathbb{1}_{X \geq d}\right) = e^{-\frac{d}{\theta}} E(X^h) = e^{-\frac{d}{\theta}}\,h!\,\theta^h. \tag{140}$$

---

1. With $E\left(N^h\right) = 1^h p + 0^h q$.

Finally, note that the moments of a gamma random variable $X \sim \mathcal{G}(\alpha, \theta)$ generalize those of the exponential random variable shown in Equations (138) and (139):

$$E(X^2) = (\alpha + 1)\,\alpha\,\theta^2 \tag{141}$$

and

$$E(X^h) = (\alpha + h - 1) \cdots \alpha\,\theta^h, \tag{142}$$

where the exponential case is again recovered by setting $\alpha = 1$.

Conditional non-centered moments can be computed as follows in the continuous case:

$$E\left(X^h \mid A\right) = \int_a^b x^h\,f(x|A)\,dx \tag{143}$$

or

$$E\left(X^h \mid A\right) = \frac{\int_a^b x^h\,f(x)\,\mathbb{1}_A\,dx}{P(A)}. \tag{144}$$

Consider the subinterval $[\alpha, \beta] \subset [a, b]$. Then, a classic application is

$$E\left(X^h \mid X \in [\alpha, \beta]\right) = \frac{\int_\alpha^\beta x^h\,f(x)\,dx}{P(X \in [\alpha, \beta])}. \tag{145}$$

In the discrete case, the expression of conditional non-centered moments is

$$E(N^h \mid A) = \sum k^h\,P(N = k|A). \tag{146}$$

The **tower property**, when it is applied to conditional non-centered moments, becomes

$$E\left(X^h\right) = E\left(E\left(X^h \mid A\right)\right). \tag{147}$$

Let $\{B_i\}$ be a countable collection of mutually exclusive events. Then, we also have:

$$E(X^h) = \sum_i E\left(X^h \mid B_i\right) P(X \in B_i), \tag{148}$$

which is an extension of the law of total probability (46).

An important remark for subsequent exams

Remember the following formulas that express the centered moments $\zeta_i = E((X - E(X))^i)$ as a function of the non-centered moments $\mu_i = E(X^i)$ at orders two, three, and four:

$$\zeta_2 = \mu_2 - \mu_1^2$$

and

$$\zeta_3 = \mu_3 - 3\,\mu_2\,\mu_1 + 2\,\mu_1^3$$

and

$$\zeta_4 = \mu_4 - 4\,\mu_3\,\mu_1 + 6\,\mu_2\,\mu_1^2 - 3\,\mu_1^4.$$

Because exams LTAM and STAM use these a lot, I suggest that you re-derive these formulas by yourself. This is an excellent exercise in the context of Exam P. In the next section, we concentrate on ζ_2 and other related measures of dispersion.

3.4 Variance and measures of dispersion

The **variance**, or the centered moment of order two[2], of a random variable X is

$$\mathrm{Var}(X) = E\left((X - E(X))^2\right). \tag{149}$$

The following equivalent formula is also very useful:

$$\mathrm{Var}(X) = E(X^2) - E(X)^2. \tag{150}$$

By extension, we can construct the **conditional variance** of X when we know the event A:

$$\mathrm{Var}(X|A) = E([X - E(X|A)]^2 | A) = E(X^2|A) - E(X|A)^2. \tag{151}$$

2. At higher orders, we also define the skewness as

$$\mathrm{Skewness}(X) = \frac{E\left((X - E(X))^3\right)}{(\mathrm{Var}(X))^{\frac{3}{2}}}$$

and the kurtosis as

$$\mathrm{Kurtosis}(X) = \frac{E\left((X - E(X))^4\right)}{(\mathrm{Var}(X))^2}.$$

The **tower property** becomes, in the context of the variance of a random variable:

$$\text{Var}(X) = E(\text{Var}(X|A)) + \text{Var}(E(X|A)). \tag{152}$$

As you will remember, in this case the variance equals the sum of the expectation of the conditional variance, and of the variance of the conditional expectation. In applications, don't forget these two terms!

Another useful measure of dispersion is the **standard deviation** that is defined as the square root of the variance:

$$\sigma_X = \sqrt{\text{Var}(X)} = \sqrt{E(X^2) - E(X)^2}. \tag{153}$$

An interesting property of σ_X is that it can be expressed in the same units as X. For instance, if X is an amount in Euros, then σ_X is also in Euros, making it directly comparable to X.

When comparing several risks, we might be interested in a measure of dispersion that is independent of the size of the risks. An example of such a measure is the **coefficient of variation** that is the ratio of the standard deviation to the mean:

$$c_v(X) = \frac{\sigma_X}{E(X)}. \tag{154}$$

The following two properties of the variance are extremely important:

$$\text{Var}(\alpha X) = \alpha^2 \text{Var}(X) \tag{155}$$

and

$$\text{Var}(X + \beta) = \text{Var}(X). \tag{156}$$

The corresponding properties for the standard deviation are

$$\sigma(\alpha X) = |\alpha|\, \sigma(X) \tag{157}$$

and

$$\sigma(X + \beta) = \sigma(X). \tag{158}$$

For a sequence of **independent identically distributed** (or **i.i.d.**) random variables $(X_i)_{i=1,\cdots,n}$, we have:

$$\text{Var}(X_1 + \cdots + X_n) = n\, \text{Var}(X_1). \tag{159}$$

Pay attention to this result and compare it to the result given in Equation (155), where you set $\alpha = n$. You see that the variance of the sum of n i.i.d. risks is different from the variance of n times a risk.

We let $\bar{X}$ denote the arithmetic average of n i.i.d. random variables:
$$\bar{X} = \frac{X_1 + \cdots + X_n}{n}.$$

Finally, you can use Equations (155) and (159) to re-derive the following result:
$$\text{Var}(\bar{X}) = \frac{\text{Var}(X_1)}{n}.$$

Discrete distributions

Let $N \sim \mathcal{B}(p)$ be a Bernoulli random variable. Its variance is
$$\text{Var}(N) = p\,q, \tag{160}$$

where here and elsewhere $q = 1 - p$.

Now, let $N \sim \mathcal{B}_{a,b}(p)$ be a generalized Bernoulli random variable that takes the following values: a with probability p and b with probability $1 - p$. Then, the variance becomes
$$\text{Var}(X) = (b-a)^2\,p\,q, \tag{161}$$

which is a very useful result for exams STAM and LTAM.

Because a binomial random variable $N \sim \mathcal{B}(n,p)$ is the sum of n i.i.d. Bernoulli random variables, its variance is equal to n times the variance of a Bernoulli random variable:
$$\text{Var}(N) = n\,p\,q, \tag{162}$$

where you observe that the variance of a binomial distribution is inferior to its mean (equal to Np).

The variance of a geometric random variable $N \sim \mathcal{G}(p)$, which represents the number of trials to reach one success, is
$$\text{Var}(N) = \frac{q}{p^2}. \tag{163}$$

The variance of a negative binomial random variable is identical for both forms of the distribution and is equal to

$$\text{Var}(N') = \text{Var}(N) = \frac{rq}{p^2}, \tag{164}$$

where the equality of the two variances is a consequence of $N' = N + r$ and of the fact that shifting a random variable by a constant has no impact on its variance.

Note that for Exam STAM, the changes of variables $p = \frac{1}{1+\beta}$ and $q = \frac{\beta}{1+\beta}$ produce the following alternative expression of the variance: $\text{Var}(N) = r\beta(1+\beta)$. This expression shows that the variance of a negative binomial distribution is superior to its mean (equal to $r\beta$).

For a Poisson random variable, the variance is equal to the mean:

$$\text{Var}(X) = E(X) = \lambda. \tag{165}$$

Finally, the variance of a hypergeometric random variable $N \sim \mathcal{H}(a, b, n)$ is

$$\text{Var}(N) = n \frac{ab}{(a+b)^2} \frac{a+b-n}{a+b-1}. \tag{166}$$

Continuous distributions

The variance of a uniform random variable $U \sim \mathcal{U}(a, b)$ is

$$\text{Var}(U) = \frac{(b-a)^2}{12}. \tag{167}$$

For a standard uniform random variable $V \sim \mathcal{U}(0, 1)$, this expression becomes

$$\text{Var}(V) = \frac{1}{12}. \tag{168}$$

The variance of an exponential random variable is equal to the square of its mean. Indeed, we have:

$$\text{Var}(X) = E(X)^2 = \theta^2. \tag{169}$$

Using Equation (140), we further obtain in the case of an exponential random variable:

$$\operatorname{Var}\left((X-d)\mathbb{1}_{X\geq d}\right) = \theta^2 \left(2 - e^{-\frac{d}{\theta}}\right) e^{-\frac{d}{\theta}}. \tag{170}$$

For a gamma random variable $X \sim \mathcal{G}(\alpha, \theta)$, the variance is equal to

$$\operatorname{Var}(X) = \alpha\,\theta^2.$$

Finally, for a normal random variable $X \sim \mathcal{N}(\mu, \sigma)$, the variance is simply equal to the square of the second parameter:

$$\operatorname{Var}(X) = \sigma^2.$$

Note that the normal distribution $\mathcal{N}(\mu, \sigma)$ can be used to approximate the Binomial distribution $\mathcal{B}(n, p)$. This is achieved by setting the means and the variances of the two distributions equal. This case yields the conditions:

$$\mu = n\,p, \tag{171}$$

and

$$\sigma^2 = n\,p\,q. \tag{172}$$

This approximation is valid when n is large and p is not too close to zero or one. We also typically require that np and $n(1-p)$ be larger than five.

Continuity correction

When approximating a discrete binomial distribution $\mathcal{B}(n, p)$ by a continuous normal distribution $\mathcal{N}(\mu, \sigma)$, an adjustment called a **continuity correction** is often made to provide more precision. This adjustment is shown below for X a normal random variable and N a binomial random variable:

$$P(N \leq k) \approx P(X \leq k + 0.5) \tag{173}$$

and

$$P(N \geq k) \approx P(X \geq k - 0.5) \tag{174}$$

and
$$P(N < k) \approx P(X < k - 0.5) \tag{175}$$
and
$$P(N > k) \approx P(X > k + 0.5). \tag{176}$$

The rationale for these approximations is as follows. In (173) and (174), we see that the values that the discrete variable N can take include the value k. In that case, it is reasonable to broaden the set of values that are admissible by the continuous approximating variable X. Conversely, in (175) and (176), we see that the values that the discrete variable N can take do *not* include the value k. In that case, it is reasonable to reduce the set of values that are admissible by the continuous approximating variable X.

Denoting a centered reduced normal random variable as $Z \sim \mathcal{N}(0,1)$, we can rewrite these four approximations as follows:

$$P(N \leq k) \approx P\left(Z \leq \frac{k + 0.5 - \mu}{\sigma}\right) = N\left(\frac{k + 0.5 - \mu}{\sigma}\right) \tag{177}$$

and

$$P(N \geq k) \approx P\left(Z \geq \frac{k - 0.5 - \mu}{\sigma}\right) = 1 - N\left(\frac{k - 0.5 - \mu}{\sigma}\right) \tag{178}$$

and

$$P(N < k) \approx P\left(Z < \frac{k - 0.5 - \mu}{\sigma}\right) = N\left(\frac{k - 0.5 - \mu}{\sigma}\right) \tag{179}$$

and

$$P(N > k) \approx P\left(Z > \frac{k + 0.5 - \mu}{\sigma}\right) = 1 - N\left(\frac{k + 0.5 - \mu}{\sigma}\right). \tag{180}$$

3.5 Sums of independent random variables

Before we look at the sum of any number of independent random variables, we consider the simple case of the sum of two random variables X and Y that are independent and that take values in $\mathbb{N}$. In that case, the distribution of the sum of X and Y is perfectly specified by the following formula:

$$P(X+Y=N) = \sum_{n=0}^{N} P(X=n)\, P(Y=N-n). \qquad (181)$$

We now come to the study of the sums of independent normal or Poisson random variables.

Sums of independent normal random variables

Let $i = 1, \cdots, n$. We consider n constants $\lambda_i \in \mathbb{R}$ and n standard normal random variables $X_i \sim \mathcal{N}(\mu_i, \sigma_i)$. We further assume that the n random variables are independent. Then, any linear combination of these variables is also normal and we have:

$$\sum_{i=1}^{n} \lambda_i X_i \sim \mathcal{N}\left(\sum_{i=1}^{n} \lambda_i \mu_i,\, \sqrt{\sum_{i=1}^{n} \lambda_i^2 \sigma_i^2}\right). \qquad (182)$$

By setting all the weights to $1/n$ in the above result, we obtain the distribution of the average of a collection of independent normal random variables:

$$\overline{X} = \frac{\sum_{i=1}^{n} X_i}{n} \sim \mathcal{N}\left(\frac{\sum_{i=1}^{n} \mu_i}{n},\, \frac{\sqrt{\sum_{i=1}^{n} \sigma_i^2}}{n}\right). \qquad (183)$$

Let us further assume that the normal random variables are identically distributed, such as $X_i \sim \mathcal{N}(\mu, \sigma)$ for all i. Then,

$$\overline{X} = \frac{\sum_{i=1}^{n} X_i}{n} \sim \mathcal{N}\left(\mu,\, \frac{\sigma}{\sqrt{n}}\right). \qquad (184)$$

By setting all the weights to one in (182), we obtain the distribution of the sum of a collection of independent normal random

variables:
$$\sum_{i=1}^{n} X_i \sim \mathcal{N}\left(\sum_{i=1}^{n} \mu_i, \sqrt{\sum_{i=1}^{n} \sigma_i^2}\right). \tag{185}$$

Let us further assume that the normal random variables are identically distributed, such as $X_i \sim \mathcal{N}(\mu, \sigma)$ for all i. Then,

$$\sum_{i=1}^{n} X_i \sim \mathcal{N}\left(\mu n, \sigma \sqrt{n}\right). \tag{186}$$

Sums of independent Poisson random variables

Let $\{N_i\}_{i=1,\cdots,n}$ denote a collection of independent Poisson random variables that have the parameters $\{\lambda_i\}_{i=1,\cdots,n}$. Then, the sum of the Poisson random variables is itself a Poisson random variable whose parameter is the sum of the parameters of the components. In a more compact way:

$$\sum_{i=1}^{n} N_i \sim \mathcal{P}\left(\sum_{i=1}^{n} \lambda_i\right). \tag{187}$$

This property can be useful for aggregating risks from various sources or over different time horizons.

3.6 Moment generating functions

The **moment generating function** of a random variable X is defined by

$$M_X(s) = E\left(e^{sX}\right), \tag{188}$$

where this function satisfies

$$M_X(0) = 1, \tag{189}$$

whatever the random variable considered.

As their name indicates, moment generating functions are useful for computing the moments of probability distributions.

The mean of X can be recovered as follows:
$$E(X) = M'_X(0). \tag{190}$$

A similar formula exists for the second non-centered moment:
$$E(X^2) = M''_X(0). \tag{191}$$

Combining the above formulas, we are able to derive a general formula for the variance:
$$\text{Var}(X) = M''_X(0) - \left(M'_X(0)\right)^2. \tag{192}$$

Equation (191) can be generalized at any order to provide an expression for the non-centered moments of a probability distribution:
$$E(X^h) = M_X^{(h)}(0), \tag{193}$$

where you note that a mere multiple differentiation of the moment generating function is required to recover non-centered moments.

For $(X_i)_{i=1,\cdots,n}$ a sequence of independent random variables, we have:
$$M_{X_1+\cdots+X_n}(s) = \prod_{i=1}^{n} M_{X_i}(s) \tag{194}$$

for all of the admissible values of s.

In the presence of a deductible d, the moment generating function associated with a claim payment is [3]:
$$M_{(X-d)^+}(s) = E\left(e^{s(X-d)^+}\right) = F(d) + \int_d^b e^{s(x-d)} f(x)\, dx. \tag{195}$$

---

3. This result is obtained by observing that
$$M_{(X-d)^+}(s) = \int_a^d e^{s \cdot 0} f(x)\, dx + \int_d^b e^{s(x-d)} f(x)\, dx,$$
where
$$\int_a^d e^{s \cdot 0} f(x)\, dx = F(d).$$

In the context of moment generating functions, the **tower property** of conditional expectations becomes:

$$M_X(s) = E\left(e^{sX}\right) = E\left(E\left(e^{sX} \mid A\right)\right) = E\left(M_X\left(s \mid A\right)\right). \quad (196)$$

Finally, we introduce another generating function called the **probability generating function**. This function is defined as follows:
$$P_X(s) = E\left(s^X\right), \quad (197)$$
for X a random variable.

When $X = N$ is a counting random variable, so when it takes values in $\mathbb{N}$, we have:

$$P_N(s) = E\left(s^N\right) = \sum_{n=0}^{+\infty} s^n \, P(N = n) \quad (198)$$

or

$$P_N(s) = P(N = 0) + s \, P(N = 1) + s^2 \, P(N = 2) + \cdots, \quad (199)$$

where we observe that

$$P_N(0) = P(N = 0). \quad (200)$$

3.7 Transformations

This section aims to characterize the **transformations** of random variables. For instance, for a given random variable X, we might want to know the probability distribution of the following transformed random variables: X^2, e^X, $\ln(X)$,...

To be able to cleanly describe the transformations of random variables, we need to refresh our knowledge about calculus and the theory of functions. Do not neglect the following paragraphs because they will help you avoid serious misinterpretations when solving exercises about transformations.

Let f be a function. Its inverse function f^{-1} is constructed such that:
$$f \circ f^{-1}(y) = y,$$

or
$$f^{-1} \circ f(x) = x.$$

We can denote each in a compact form as
$$f^{-1} \circ f = f \circ f^{-1} = \text{Id},$$

where Id is the identity function that associates x with x.

The inverse function f^{-1} exists whenever f is a **bijection**. This is the case when all of the elements in the range (set of images) of f admit a unique antecedent in the domain (set of starting points) of f. This definition guarantees that all of the elements in the domain of f^{-1} admit a unique image, so that indeed f^{-1} is well-defined.

In practice, the function f must be **strictly increasing** (in which case f^{-1} will also be strictly increasing) or **strictly decreasing** (then f^{-1} will also be strictly decreasing).

Take the example of the exponential function. Its domain of definition is $\mathbb{R}$ and it produces images in the range $\mathbb{R}^{+*}$. Consider any element of $\mathbb{R}^{+*}$: it admits a unique antecedent in $\mathbb{R}$ via the exponential function. The exponential function is therefore a bijection. Thus, we can construct its inverse, which is the natural logarithm whose domain of definition is $\mathbb{R}^{+*}$ and whose images pertain to $\mathbb{R}$. Just like the exponential, the logarithm is a strictly increasing function.

We are now equipped to study the transformations of random variables. Let $Y = f(X)$, where f is a strictly increasing function. Then,

$$P(Y \leq y) = P(f(X) \leq y) = P(X \leq f^{-1}(y)), \qquad (201)$$

where f^{-1} is the inverse function of f.

Let $Y = f(X)$, where f is a strictly decreasing function. Then,

$$P(Y \leq y) = P(f(X) \leq y) = P(X \geq f^{-1}(y)) = 1 - P(X \leq f^{-1}(y)), \qquad (202)$$

where the strict decrease of f^{-1} entails the inversion of the inequality sign in the second equality.

There are situations where f is neither increasing nor decreasing on its whole domain of definition. In such a situation, a case-by-case approach is necessary, such as with

$$P(Y = X^2 \leq y) = P(-\sqrt{y} \leq X \leq \sqrt{y}), \qquad (203)$$

where the function $Y = f(X) = X^2$ is increasing for positive values of X and decreasing for negative values of X.

You will often be asked to derive the density of a transformed random variable and you will be given the relation $Y = f(X)$ together with the density or the cdf of X. A typical way of obtaining the answer is as follows:

— If necessary, obtain the cumulative distribution function of X from its density using Equation (68).
— Derive the cdf of Y using Equation (201) or Equation (202).
— Differentiate the cdf of Y to obtain the density of Y, as in Equation (70).

Chapter 4

Multivariate probability distributions

We now come to the study of multivariate probability functions and dependent random variables $X_{i=1,\cdots,n}$, where

$$P(X_1 \leq x_1, \cdots, X_n \leq x_n) \neq \prod_{i=1}^{n} P(X_i \leq x_i).$$

4.1 Joint and marginal probability functions and densities

We first study the continuous setting and a few classic illustrations. Then, we move on to the discrete setting.

Continuous setting

In the continuous setting, a **joint probability density function** $f_{X,Y}$ is defined as follows:

$$f_{X,Y}(x,y)\, dx\, dy = P(x \leq X < x+dx,\ y \leq Y < y+dy),$$

where dx and dy are infinitesimal increments around x and y, respectively.

It is possible to recover a **marginal density function** (also called an individual density function) from a joint density function by computing:

$$f_X(x) = \int_{\mathcal{D}(y)} f_{X,Y}(x,y)\,dy, \qquad (204)$$

where $\mathcal{D}(y)$ is the set of values taken by the random variable Y.

The factorization property that characterizes the independence of events or random variables also exists in the case of densities:

$$f(x,y) = g(x)\,h(y) \quad \forall (x,y) \in \mathcal{D}(x,y) \quad \Leftrightarrow \quad X \perp Y, \qquad (205)$$

where $\mathcal{D}(x,y)$ is the set of values taken by the random variables (X,Y) and where $X \perp Y$ is the notation for the independence of X and Y.

Be careful to incorporate potential inequality conditions in the definitions of f, g and h. For example:

$$f(x,y) = \frac{8}{3}\,x\,y, \quad 0 \le x \le 1,\ x \le y \le 2x$$

has a form in $x\,y$ that could lead you to conclude that X and Y are independent but this would be wrong. Indeed, you must rewrite:

$$f(x,y) = \frac{8}{3}\,x\,y\,\mathbb{1}_{0 \le x \le 1}\,\mathbb{1}_{x \le y \le 2x},$$

which cannot be expressed as the product of a function of only x and a function of only y.

Assume you are given several inequality constraints on x and y and you want to compute a probability expression (probability, mean, variance...) involving the joint density function on the domain corresponding to the set of inequalities. Two main approaches exist to solve that kind of problem. You can either plot the various inequalities as in the illustrative Figure 4, where the gray zone is the set of points that satisfies all the inequalities given in the graph.

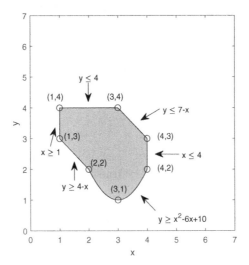

Figure 4 – Graphical Representation of a Series of Inequalities

The second approach amounts to transforming the collection of inequalities into two inequalities written as follows:

$$\begin{cases} a < x < b \\ c(x) < y < d(x). \end{cases} \qquad (206)$$

The searched probability quantity is then computed by setting the integral with respect to y as the inner integral to be computed first and by setting the integral with respect to x as the outer integral to be computed in a second step. For instance,

$$E(g(X,Y)) = \int_a^b \left[\int_{c(x)}^{d(x)} g(x,y) \, f(x,y) \, dy \right] dx, \qquad (207)$$

where f is the joint density of X and Y and g is any function that allows the integrals to converge.

Probabilities are derived from Equation (207) by using indicator functions. For instance, the cumulative distribution function

Probability Theory

of $X+Y$ is obtained by setting $g(X,Y) = \mathbb{1}_{X+Y\leq z}$, so that

$$P(X+Y \leq z) = E(\mathbb{1}_{X+Y\leq z}). \tag{208}$$

Similarly,
$$P(X \leq Y) = E(\mathbb{1}_{X\leq Y}), \tag{209}$$
and so on for any probabilistic quantity involving X and Y.

As it is very easy to make mistakes with these exercises, I suggest that you always perform both the graphical approach and the analytical approach to check one approach against the other.

Classic illustrations

Let $Z = (X, Y)$ be a uniform random variable defined on a two-dimensional domain D. Its density is

$$f(x,y) = \frac{1}{\mu(D)} \mathbb{1}_{(x,y)\in D}, \tag{210}$$

where $\mu(D)$ is the surface of domain D.

Let $X \sim \mathcal{N}(\mu_X, \sigma_X)$ and $Y \sim \mathcal{N}(\mu_Y, \sigma_Y)$ be two correlated normal random variables with a correlation coefficient of ρ (*the clean mathematical definition of a correlation will be given in a subsequent section*). We also write $(X,Y) \sim \mathcal{N}(\mu_X, \sigma_X, \mu_Y, \sigma_Y, \rho)$. The joint density function of these two normal random variables is given by

$$f_{X,Y}(x,y) = \frac{1}{2\pi\sigma_X\sigma_Y\sqrt{1-\rho^2}} e^{-\frac{\frac{(x-\mu_X)^2}{\sigma_X^2} + \frac{(y-\mu_Y)^2}{\sigma_Y^2} - \frac{2\rho(x-\mu_X)(y-\mu_Y)}{\sigma_X\sigma_Y}}{2(1-\rho^2)}}.$$

See how this form generalizes the form given in Equation (64). Then, set ρ equal to zero in the above equation. In that case, we recover

$$f_{X,Y}(x,y) = f_X(x)\, f_Y(y).$$

This equality shows that two uncorrelated normal random variables are independent. We can formalize this result as follows:

$$\rho = \rho(X,Y) = 0 \quad \Rightarrow \quad X \perp Y, \tag{211}$$

when $(X, Y) \sim \mathcal{N}(\mu_X, \sigma_X, \mu_Y, \sigma_Y, \rho)$.

Note that this result does not necessarily hold in the general case of non-Gaussian random variables, where uncorrelated random variables *can* be dependent.

However, the converse result is always true: independent random variables always have a null correlation.

Discrete setting

In the discrete setting, a **joint probability mass function** is defined as follows:

$$f(x, y) = P(X = x, Y = y).$$

Note that Equation (207) has the following discrete equivalent form:

$$E(g(M, N)) = \sum_{j=j_{\min}}^{j_{\max}} \sum_{k=k_{\min}(j)}^{k_{\max}(j)} g(j, k) \, P(M = j, N = k). \quad (212)$$

Also note that in the discrete case we can recover **marginal probabilities** as follows:

$$P(X = k) = \sum_{i=1}^{i_{\max}} P(X = k, Y = i). \quad (213)$$

Now, independence is again characterized by the factorization of probabilities:

$$X \perp Y \quad \Rightarrow \quad \forall (k, i) \quad P(X = k, Y = i) = P(X = k) \, P(Y = i). \quad (214)$$

4.2 Joint cumulative distribution functions

In the **discrete case**, a **joint cumulative distribution function** is defined by

$$F_{M,N}(m, n) = P(M \leq m, N \leq n) = \sum_{j=j_{\min}}^{m} \sum_{k=k_{\min}}^{n} P(M = j, N = k). \quad (215)$$

We can recover the total probability by setting:

$$F_{M,N}(j_{\max}, k_{\max}) = \sum_{j=j_{\min}}^{j_{\max}} \sum_{k=k_{\min}}^{k_{\max}} P(M=j, N=k) = 1. \quad (216)$$

In the **continuous case**, a **joint cumulative distribution function** is defined by

$$F_{X,Y}(x,y) = P(X \leq x, Y \leq y) = \int_{a_1}^{x} \int_{a_2}^{y} f_{X,Y}(s,t) \, ds \, dt, \quad (217)$$

where X takes its values in $[a_1, b_1]$, Y takes its values in $[a_2, b_2]$, and a_1 and a_2 may be replaced with functions of s or t to account for constraints between X and Y.

We can recover the total probability by setting:

$$F_{X,Y}(b_1, b_2) = P(X \leq b_1, Y \leq b_2) = 1. \quad (218)$$

We can also define a **joint survival function** as follows:

$$S_{X,Y}(x,y) = P(X > x, Y > y) = \int_{x}^{b_1} \int_{y}^{b_2} f_{X,Y}(s,t) \, ds \, dt, \quad (219)$$

where b_1 and b_2 may be replaced with functions of s or t to account for constraints between X and Y.

In full generality, $S_{X,Y}(x,y) \neq 1 - F_{X,Y}(x,y)$ because of standard results on sets (the complement of $\{X \leq x, Y \leq y\}$ is $\{X > x\} \cup \{Y > y\}$ but *not* $\{X > x\} \cap \{Y > y\} = \{X > x, Y > y\}$).

Further, and as already hinted in the beginning of the chapter, $F_{X,Y}$ can be factored out as the product of F_X and F_Y when X and Y are independent:

$$X \perp Y \quad \Leftrightarrow \quad F_{X,Y}(x,y) = F_X(x) \, F_Y(y). \quad (220)$$

It is possible to come back to a marginal cumulative distribution function with

$$F_X(x) = F_{X,Y}(x, b_2), \quad (221)$$

where b_2 is the maximum value attainable by Y.

Finally, joint probabilities can be recovered from joint cumulative distribution functions as follows in the discrete case:

$$P(M = m, N = n) = F(m,n) - F(m-1,n) - F(m,n-1) + F(m-1,n-1). \tag{222}$$

Whereas, in the continuous case:

$$f_{X,Y}(s,t) = \left(\frac{\partial^2 F_{X,Y}(x,y)}{\partial x \partial y} \right)_{x=s, y=t}. \tag{223}$$

4.3 Central limit theorem

Let $(X_i)_{i=1,\cdots,n}$ be a sequence of independent identically distributed random variables. The distribution of these random variables is *a priori* unknown. We only know that they have a common finite mean μ and a common finite standard deviation σ. Then, asymptotically (for n large - and in a sense that is currently out of the scope of this book), their sum is distributed as follows:

$$\sum_{i=1}^{n} X_i \sim \mathcal{N}\left(n\mu, \sqrt{n}\sigma\right). \tag{224}$$

This is the well-known **Central Limit Theorem**. This result is extremely powerful in practice: it tells you that you can aggregate nearly any type of distributions and end up with the simple normal distribution. Note that this theorem may not hold for exotic distributions for which the standard deviation is not finite, but that situation is rarely encountered.

4.4 Conditional probability distributions

In the **continuous case**, the **conditional density** of X when you know Y is given by

$$f_{Y|X}(y|x) = \frac{f_{X,Y}(x,y)}{f_X(x)}. \tag{225}$$

This is a function of both x and y. Using Equation (204), we can obtain the alternative representation:

$$f_{Y|X}(y|x) = \frac{f_{X,Y}(x,y)}{\int_{\mathcal{D}(y)} f_{X,Y}(x,y)\,dy}, \qquad (226)$$

where $\mathcal{D}(y)$ is the set of values for the random variable Y.

For $[a,b] \subset \mathcal{D}(y)$, a classic application of the above formula is

$$P(a \leq Y \leq b \mid X = x) = \int_a^b f(y|x)\,dy = \frac{\int_a^b f(x,y)\,dy}{\int_{\mathcal{D}(y)} f(x,y)\,dy}, \qquad (227)$$

which is a function of x only.

In the **discrete case**, a formula analogous to Equation (226) is

$$P(Y = i \mid X = k) = \frac{P(X = k, Y = i)}{P(X = k)} = \frac{P(X = k, Y = i)}{\sum_{i=1}^{i_{\max}} P(X = k, Y = i)}. \qquad (228)$$

4.5 Moments

Moments of probability distributions can be viewed as subcases of the general formula (207). We rewrite it here in a simplified notation:

$$E(g(X,Y)) = \int\int g(x,y)\,f(x,y)\,dx\,dy, \qquad (229)$$

where g is any regular enough function and f is the joint density of X and Y.

As an example, the first order cross-moment of X and Y can be written as a function of f:

$$E(XY) = \int\int x\,y\,f(x,y)\,dx\,dy. \qquad (230)$$

Similarly, the moment of order h of Y also depends on the function f in the following way:

$$E(Y^h) = \int \int y^h \, f(x,y) \, dx \, dy. \tag{231}$$

In practice, integrations can be performed in any order such as

$$E(Y) = \int \left(\int y \, f(x,y) \, dx \right) dy = \int \left(\int y \, f(x,y) \, dy \right) dx. \tag{232}$$

However, you should choose any of the two orders of integration with care depending on the problem you are facing.

Conditional Moments

A **conditional moment**, or conditional expectation, should consistently be written as a function of the conditional density:

$$E\left(Y^h \mid X = x\right) = \int y^h \, f(y|x) \, dy. \tag{233}$$

Be careful not to confound the conditional expectation of Y *when* you know X:

$$E\left(Y \mid X = x\right) = \int y \, f(y|x) \, dy, \tag{234}$$

which depends on the conditional density $f(y|x)$, and the 'joint' expectation of Y *and* the knowledge of X:

$$E\left(Y \, ; \, X = x\right) = \int y \, f(x,y) \, dy, \tag{235}$$

which depends on the joint density $f(x,y)$.

Note that

$$E\left(Y^h \mid X = x\right) = \frac{\int y^h \, f(x,y) \, dy}{f(x)} = \frac{\int y^h f(x,y) \, dy}{\int f(x,y) \, dy}, \tag{236}$$

where the first equality follows from (233) and the definition of a conditional density and the second equality follows from (204).

Note that the first equality in (236) can be rewritten as follows:
$$E\left(Y^h \mid X = x\right) = \frac{E\left(Y^h \; ; \; X = x\right)}{f(x)}, \qquad (237)$$
which provides a nice interpretation of the link between a conditional expectation and a 'joint' expectation.

Tower property

In the presence of two risks X and Y, the tower property of conditional expectations becomes
$$E(Y) = E(E(Y|X)), \qquad (238)$$
where in the right-hand side of the equation we first compute the inner expectation with respect to the conditional distribution of Y when you know X. The outer expectation is computed with respect to the distribution of X in a second step.

More generally, we have:
$$E(g(X,Y)) = E(E(g(X,Y)|X)). \qquad (239)$$

In the particular case where X and Y are independent, we can compute $E(g(X)h(Y))$ as follows:
$$X \perp Y \quad \Rightarrow \quad E(g(X).h(Y)) = E(g(X))\, E(h(Y)). \qquad (240)$$

But, in general this factorization cannot be performed. Therefore, we need to use the tower property of conditional expectations. By applying Equation (239), we obtain
$$E(g(X).h(Y)) = E\left[g(X)\, E(h(Y)|X)\right], \qquad (241)$$
where $g(X)$ can be factored out of the operator $E(.|X)$ by definition of this latter quantity.

Similarly, we can derive
$$E(X.Y) = E(X.E(Y|X)). \qquad (242)$$

By combining Equations (238) and (234), we also obtain

$$E(Y) = \int \left(\int y\, f(y|x)\, dy \right) f(x)\, dx. \qquad (243)$$

Finally note that it is possible to rewrite Equation (238) in terms of a joint expectation rather than in terms of a conditional expectation. In fact,

$$E(Y) = \int E(Y; X)\, dx. \qquad (244)$$

As an exercise, you can derive this formula by plugging Equation (237), with $h = 1$, into Equation (238). Then, you need to check that the density term $f(x)$ does indeed disappear.

Discrete case

Conditional moments can be written as follows in the discrete case:

$$E\left(Y^h \mid X = k\right) = \sum_{i=1}^{i_{\max}} i^h P(Y = i \mid X = k). \qquad (245)$$

An expression in terms of joint moments also exists:

$$E\left(Y^h \mid X = k\right) = \frac{E\left(Y^h\,;\, X = k\right)}{P(X = k)} = \frac{\sum_{i=1}^{i_{\max}} i^h P(X = k, Y = i)}{\sum_{i=1}^{i_{\max}} P(X = k, Y = i)}. \qquad (246)$$

Similarly,

$$E\left(Y^h \mid X \leq k\right) = \frac{E\left(Y^h\,;\, X \leq k\right)}{P(X \leq k)} = \frac{\sum_{i=1}^{i_{\max}} i^h P(X \leq k, Y = i)}{\sum_{i=1}^{i_{\max}} P(X \leq k, Y = i)}. \qquad (247)$$

A useful application of the tower property (238) is given by

$$E\left(Y^h\right) = \sum_k E\left(Y^h; X = k\right) = \sum_k E\left(Y^h \mid X = k\right) P(X = k),$$
(248)

where in this expression Y is often assumed to be one of the following: $Y = Z \ \mathbb{1}_{Z \geq \alpha}$ or $Y = G^2 \mid H \leq \beta$.

Random parameter

The tower property can be used to deal with the case of a random parameter. Assume that the distribution of Y depends on parameter X, which is a random variable. Then,[1]

$$P(Y = k) = E(P(Y = k \mid X)) = \int P(Y = k \mid x) \, f(x) \, dx,$$
(249)

where f is the density of the distribution of X.

This situation is tested once in Exam P and will be encountered often in Exam STAM.

Collective model

Let $(H_i)_{i=1,\ldots,+\infty}$ be a collection of independent identically distributed random variables that are also independent from another random variable N. The sequence $(H_i)_{i=1,\ldots,+\infty}$ typically represents the **claims** of an insured entity or of a pool of insured entities. The random variable N counts the **number of claims**. Let

$$X = \sum_{i=1}^{N} H_i$$

be the **aggregate claim** of the insured entity or pool of insured entities.

---

1. Use Equation (238) with $P(Y = k) = E\left(\mathbb{1}_{Y=k}\right)$ and $P(Y = k \mid X) = E\left(\mathbb{1}_{Y=k} \mid X\right)$.

Then, using the tower property of conditional expectations [2], we can compute the expected aggregate claim as follows:

$$E(X) = E(N)\, E(H_1). \tag{250}$$

This result is very useful to know for Exam STAM as well.

Normal conditional on normal

Let $(X, Y) \sim \mathcal{N}(\mu_X, \sigma_X, \mu_Y, \sigma_Y, \rho)$. Then, the conditional expectation of the normal random variable Y when you know the value of the normal random variable X is

$$E(Y \mid X = x) = E(Y) + \frac{\rho\, \sigma_Y}{\sigma_X}(x - E(X)), \tag{251}$$

which is linear in x.

This result has applications in insurance, portfolio management, derivatives pricing, and a lot of other fields.

4.6 Joint moment generating functions

The **joint moment generating function** for the random variables X and Y is defined by

$$M_{X,Y}(u, v) = E\left(e^{uX + vY}\right). \tag{252}$$

When X and Y are independent, this result simplifies to:

$$M_{X,Y}(u, v) = E\left(e^{uX}\right) E\left(e^{vY}\right). \tag{253}$$

---

2. We compute

$$E(X) = E\left(\sum_{i=1}^{N} H_i\right) = E\left(E\left(\sum_{i=1}^{N} H_i \mid N\right)\right) = E\left(E(N\, H_1 \mid N)\right),$$

where the second equality is a consequence of the tower property of conditional expectations and where the third equality is a consequence of the i.i.d property of the $(H_i)_{i=1,\ldots,N}$ sequence, noting that N is a constant within the inner conditional expectation operator $E(\,.\mid N)$. Then, we obtain

$$E(X) = E\left(N\, E(H_1)\right) = E(N)\, E(H_1),$$

where in the first equality we simply factor out N from the inner conditional expectation and in the second equality we use the fact that $E(H_1)$ is a constant from the viewpoint of the exterior expectation.

4.7 Variance and measures of dispersion

The **conditional variance** of Y when you know X is defined by

$$\text{Var}(Y \mid X) = E\left([Y - E(Y|X)]^2 \mid X\right) = E\left(Y^2|X\right) - E\left(Y|X\right)^2. \tag{254}$$

The **tower property** of conditional variances, also called the **law of total variance**, is given by

$$\text{Var}(Y) = E(\text{Var}(Y|X)) + \text{Var}(E(Y|X)). \tag{255}$$

Note that this equation is a direct adaptation of Equation (152).

Collective model

Let $(H_i)_{i=1,\ldots,+\infty}$ be a collection of independent identically distributed random variables that are also independent from the random variable N. Again, each H_i can be interpreted as a claim and N can be interpreted as a number of claims. Define

$$X = \sum_{i=1}^{N} H_i.$$

Then, using the tower property of conditional expectations, we obtain [3]

$$\text{Var}(X) = E(N)\,\text{Var}(H_1) + \text{Var}(N)\,E(H_1)^2. \tag{256}$$

---

3. The tower property of conditional variance allows us to write:

$$\text{Var}(X) = \text{Var}\left(\sum_{i=1}^{N} H_i\right) = E\left(\text{Var}\left(\sum_{i=1}^{N} H_i | N\right)\right) + \text{Var}\left(E\left(\sum_{i=1}^{N} H_i | N\right)\right).$$

We first compute

$$E\left(\text{Var}\left(\sum_{i=1}^{N} H_i \mid N\right)\right) = E\left(N\,\text{Var}\left(H_1\right)\right) = E\left(N\right)\,\text{Var}\left(H_1\right),$$

where the first equality is a consequence of the i.i.d property of the $(H_i)_{i=1,\ldots,N}$ sequence - noting that N is a constant within the inner conditional variance - and of Equation (159). In the second equality, we use the

This result will also be very useful to know for Exam STAM.

Normal conditional on normal

Let $(X, Y) \sim \mathcal{N}(\mu_X, \sigma_X, \mu_Y, \sigma_Y, \rho)$. Then, the conditional variance of Y when you know X is

$$\text{Var}(Y \mid X = x) = (1 - \rho^2) \, \text{Var}(Y). \tag{257}$$

4.8 Covariance and correlation coefficients

The **covariance** between the random variables X and Y is defined by

$$\text{Cov}(X, Y) = E\left[(X - E(X))(Y - E(Y))\right], \tag{258}$$

which can be written equivalently as

$$\text{Cov}(X, Y) = E(XY) - E(X) \, E(Y). \tag{259}$$

This operator is not easy to interpret because it measures not only the dependence between X and Y but also the sizes of X and Y. A dependence measure that does not depend on the sizes of X and Y and that is unitless is the **correlation coefficient** defined by

$$\text{Corr}(X, Y) = \frac{\text{Cov}(X, Y)}{\sigma_X \sigma_Y}. \tag{260}$$

The correlation coefficient varies between -1 and 1. It is equal to 1 in the case of full dependence and it is equal to -1 if the random variables always move in opposite directions.

---

fact that $\text{Var}(H_1)$ is a constant from the viewpoint of the exterior expectation. Then, we have:

$$\text{Var}\left(E\left(\sum_{i=1}^{N} H_i \mid N\right)\right) = \text{Var}\left(N E(H_1)\right) = \text{Var}(N) \, E(H_1)^2,$$

where the first equality is again a consequence of the i.i.d property of the $(H_i)_{i=1,\ldots,N}$ sequence, noting that N is a constant within the inner conditional expectation. In the second equality, we use the fact that $E(H_1)$ is a constant from the viewpoint of the exterior variance operator.

Let X and Y be two correlated random variables whose standard deviations are σ_X and σ_Y and whose correlation coefficient is ρ. Then,

$$\operatorname{Var}(X+Y) = \operatorname{Var}(X) + \operatorname{Var}(Y) + 2\operatorname{Cov}(X,Y), \qquad (261)$$

or

$$\operatorname{Var}(X+Y) = \sigma_X^2 + \sigma_Y^2 + 2\,\rho\,\sigma_X\sigma_Y. \qquad (262)$$

This latter quantity can be conveniently extended as follows:

$$\operatorname{Var}(\alpha X + \beta Y) = \alpha^2 \sigma_X^2 + \beta^2 \sigma_Y^2 + 2\,\alpha\beta\,\rho\,\sigma_X\sigma_Y, \qquad (263)$$

for any two constants α and β. This equation is equivalent to

$$\operatorname{Var}(\alpha X + \beta Y) = \alpha^2 \operatorname{Var}(X) + \beta^2 \operatorname{Var}(Y) + 2\,\alpha\beta\,\operatorname{Cov}(X,Y). \qquad (264)$$

We also have:

$$\operatorname{Var}(X-Y) = \operatorname{Var}(X) + \operatorname{Var}(Y) - 2\operatorname{Cov}(X,Y) \qquad (265)$$

or

$$\operatorname{Var}(X-Y) = \sigma_X^2 + \sigma_Y^2 - 2\rho\sigma_X\sigma_Y, \qquad (266)$$

which can be conveniently extended as follows:

$$\operatorname{Var}(\alpha X - \beta Y) = \alpha^2 \sigma_X^2 + \beta^2 \sigma_Y^2 - 2\,\alpha\beta\,\rho\,\sigma_X\sigma_Y. \qquad (267)$$

Note that Equation (267) can be readily obtained from Equation (263) by setting $\beta := -\beta$.

Consider now the following subcase. Let $X \sim \mathcal{N}(\mu_1, \sigma_1)$ and $Y \sim \mathcal{N}(\mu_2, \sigma_2)$ be two normal random variables that have a correlation coefficient equal to ρ. Then,

$$X + Y \sim \mathcal{N}\left(\mu_1 + \mu_2,\, \sqrt{\sigma_1^2 + \sigma_2^2 + 2\rho\sigma_1\sigma_2}\right). \qquad (268)$$

Finally note that the conditional covariance of X and Y when you know Z is defined by

$$\operatorname{Cov}(X, Y \mid Z) = E\left([X - E(X \mid Z)][Y - E(Y \mid Z)] \mid Z\right), \qquad (269)$$

or, equivalently, by

$$\operatorname{Cov}(X, Y \mid Z) = E(XY \mid Z) - E(X \mid Z) E(Y \mid Z). \quad (270)$$

The **tower property** of conditional covariances, also called the **law of total covariance**, is given by

$$\operatorname{Cov}(X, Y) = E(\operatorname{Cov}(X, Y|Z)) + \operatorname{Cov}(E(X|Z), E(Y|Z)). \quad (271)$$

From dependent to independent normal random variables

Let $(X, Y) \sim \mathcal{N}(\mu_X, \sigma_X, \mu_Y, \sigma_Y, \rho)$ be a couple of dependent normal random variables. The following two random variables are normal and independent:

$$X \quad \perp \quad Z = Y - \frac{\rho \, \sigma_Y}{\sigma_X} X, \quad (272)$$

where Z is normal because it is a linear combination of normal random variables and where the independence occurs because $\operatorname{Cov}(X, Z) = 0$.

In greater detail, we have:

$$(X, Z) \sim \mathcal{N}\left(\mu_X, \sigma_X, \mu_Y - \frac{\rho \, \sigma_Y}{\sigma_X} \mu_X, \sigma_Y \sqrt{1 - \rho^2}, 0\right), \quad (273)$$

where the computation of μ_Z and σ_Z is left as an exercise for the reader.

It is often useful to decompose a random variable as a sum of two independent components. Here, we can decompose Y as follows:

$$Y = Z + (Y - Z) = \left(Y - \frac{\rho \, \sigma_Y}{\sigma_X} X\right) + \frac{\rho \, \sigma_Y}{\sigma_X} X, \quad (274)$$

where the last equality above and the result in (272) show that the two components Z and $Y - Z$ of Y are indeed independent.

Finally, we can obtain two independent standard normal random variables by defining

$$(V, W) = \left(\frac{X - \mu_X}{\sigma_X}, \frac{Z - \mu_Z}{\sigma_Z}\right), \quad (275)$$

so that

$$(V, W) = \left(\frac{X - \mu_X}{\sigma_X}, \frac{Y - \mu_Y - \frac{\rho \sigma_Y}{\sigma_X}(X - \mu_X)}{\sigma_Y \sqrt{1 - \rho^2}} \right). \tag{276}$$

These new normal random variables satisfy

$$(V, W) \sim \mathcal{N}(0, 1, 0, 1, 0). \tag{277}$$

4.9 Transformations and order statistics

Let $(X_i)_{i=1,\cdots,n}$ be a collection of random variables that do not satisfy any particular constraint. Let us also denote their maximum by $M = \max(X_1, \cdots, X_n)$, while their minimum is $m = \min(X_1, \cdots, X_n)$. Then,

$$P(M \leq x) = P\left(\bigcap_{i=1}^{n} (X_i \leq x) \right) \tag{278}$$

and

$$P(m > x) = P\left(\bigcap_{i=1}^{n} (X_i > x) \right). \tag{279}$$

We now add the constraint that the $(X_i)_{i=1,\cdots,n}$ are independent random variables. We have:

$$P(M \leq x) = \prod_{i=1}^{n} P(X_i \leq x) \tag{280}$$

and

$$P(m > x) = \prod_{i=1}^{n} P(X_i > x). \tag{281}$$

Further assume that the $(X_i)_{i=1,\cdots,n}$ are identically distributed. Then,

$$P(M \leq x) = P(X_1 \leq x)^n \tag{282}$$

and

$$P(m > x) = P(X_1 > x)^n. \tag{283}$$

Appendix

The tables provided in this appendix give hints for solving the series of exercises made freely available by the Society of Actuaries. They can be read as follows: Ex. 1, or Exercise 1, is solved using Eq. 20, or Equation 20, from this book. T1 is the first table and F4 the fourth figure of this book. / means the exercise has been suppressed by the SOA.

Ex.	Eq.	Ex.	Eq.	Ex.	Eq.
1	20	26	15, 48	51	104
2	6, 19	27	48	52	66, 102
3	11, 19	28	15, 48, 51	53	107
4	19, 30	29	84	54	104
5	16	30	56	55	69, 103
6	41	31	76	56	104
7	45	32	52	57	192
8	6, 19, 15	33	42, 72	58	193, 194
9	6, 19, 41	34	68, 69	59	96
10	/	35	/	60	155
11	15, 30	36	42, 72	61	/
12	16, 41	37	71, 84	62	131, 150
13	8, 15, 41	38	41, 68, 71	63	131, 150
14	25	39	76	64	130, 150
15	8, 19, 20	40	68	65	132, 150
16	181	41	77	66	/
17	34	42	89	67	56, 153
18	92, 88, 182	43	54	68	61, 69, 96
19	48	44	102	69	84, 85, 96
20	48	45	103	70	41, 84, 96
21	43, 48	46	103	71	70
22	48	47	103	72	82
23	48	48	102	73	70, 201
24	41	49	/	74	70, 202
25	48	50	56, 102	75	70, 201

Table 2 – Exercises and their related equations, tables, and figures.

Ex.	Eq.	Ex.	Eq.	Ex.	Eq.
76	70, 282	101	156, 263	126	66
77	15, 219	102	159, 169	127	71, 132, 224
78	/	103	15, 84, 278	128	20, 16
79	15, 219	104	205, 259	129	41, 84, 87
80	T1, 99, 224	105	F4, 206, 207	130	188
81	89, 183	106	238, 242, 259	131	112, 228
82	71, 165, 224	107	259, 261	132	57
83	89, 186	108	206, 207, 208	133	41, 71
84	88, 224, 262	109	206, 207	134	23, 25
85	89, 169, 224	110	227	135	152, 165
86	88, 224	111	227	136	112, 248
87	92, 167, 224	112	227	137	194
88	30, 84	113	16, 41	138	239
89	206, 207	114	246	139	247
90	120, 206, 207	115	236, 254	140	55
91	F4, 206, 207	116	246, 254	141	36
92	F4, 206, 207	117	68, 204	142	248
93	F4, 206, 207	118	204, 206	143	6, 19, 15
94	F4, 206, 207	119	68, 204, 225	144	61, 206, 217
95	252, 253	120	68, 204, 225	145	236, 254
96	51	121	232	146	20, 18
97	F4, 206, 207	122	232	147	170
98	188	123	16, 87	148	165, 169, 256
99	157, 158, 262	124	143, 151	149	162, 169, 256
100	45, 130, 150	125	204, 225	150	41, 96

Table 3 – Exercises and their related equations, tables, and figures.

Ex.	Eq.	Ex.	Eq.	Ex.	Eq.
151	57	176	47	201	48
152	46, 53, 81	177	57	202	16, 41
153	115, 137	178	131, 150	203	226, 233
154	3, 24	179	5, 19, 15	204	225, 207, 208
155	59, 130, 131	180	82, 93	205	82
156	24, 41	181	43, 47	206	59, 132
157	103	182	44, 48	207	11
158	123	183	70, 203	208	46
159	24	184	24	209	105
160	262	185	48	210	37, 38, 65
161	108	186	89, 172, 178	211	57
162	230, 231, 264	187	56, 65	212	78, 187
163	171, 172, 177	188	24	213	78, 115
164	56, 187	189	96, 99, 100	214	85
165	150, 193	190	182, 96, 100	215	68, 69, 72
166	130, 150	191	257	216	56, 66, 102
167	42, 68, 69	192	82	217	59, 107
168	32, 33	193	61, 85	218	96, 99
169	46	194	95	219	99, 100
170	58	195	95	220	145
171	204	196	41, 51, 53	221	88, 99, 100
172	46, F4	197	41, 55	222	153, 193, 192
173	78, 187	198	24	223	182, 183, 91
174	77	199	24, 87, 120	224	207, 218
175	71, 88, 89	200	57	225	212, 216

Table 4 – Exercises and their related equations, tables, and figures.

Ex.	Eq.	Ex.	Eq.	Ex.	Eq.
226	102, 128	251	57, 76	276	121
227	150, 212, 216	252	116, 78	277	121
228	204, 226, 234	253	20	278	56, 102, 115
229	47, 51, 126	254	23	279	56, 74
230	70, 103, 221	255	36, 46	280	102, 127
231	24, 146, 151	256	42	281	59, 107
232	70, 131, 221	257	19, 41	282	59, 96, 129
233	169, 204	258	58	283	42, 96, 99
234	46, 130, 150	259	28	284	42, 99, 100
235	89, 186, 265	260	48	285	96, 99
236	24, 95	261	43, 48	286	68, 69
237	162, 264	262	46, 51, 56	287	84, 169
238	42, 51, 128	263	51, 53	288	104, 132, 150
239	29	264	51	289	137, 148, 150
240	41, 88, 101	265	29, 37	290	56, 148, 150
241	96	266	87	291	59, 134, 150
242	88, 251, 257	267	87	292	132, 150, 167
243	24	268	88, 100	293	88, 96, 100
244	24	269	75	294	153
245	102	270	49, 84, 124	295	57, 58, 246
246	20	271	42, 84	296	59, 195
247	37, 39	272	21, 31	297	189, 194
248	41, 46, 57	273	24, 41, 43	298	56
249	24, 51	274	42, 81	299	67, 87
250	22, 30	275	42, 51, 76	300	61, 207, 209

Table 5 – Exercises and their related equations, tables, and figures.

Ex.	Eq.	Ex.	Eq.
301	34, 59	326	137, 248
302	5, 19, 221	327	102
303	24, 76, 220	328	198, 200
304	222	329	56, 199
305	96, 99, 162	330	84, 87
306	89, 159, 161	331	98, 118, 167
307	56, 59, 249	332	41, 204, 219
308	135, 150, 240		
309	207		
310	24, 102		
311	210, 232		
312	102, 213		
313	51, 196		
314	59, 167, 204		
315	152, 167, 169		
316	245, 254		
317	167, 169, 255		
318	130, 150, 213		
319	59, 207, 270		
320	214, 259		
321	84, 281		
322	88, 182		
323	89, 182		
324	154, 261		
325	156, 261		

Table 6 – Exercises and their related equations, tables, and figures.

FINANCIAL MATHEMATICS

KEY CONCEPTS AND TOOLS for
SOA EXAM FM & CAS EXAM 2

OLIVIER LE COURTOIS

Made in the USA
Middletown, DE
04 December 2019